Robert William Burnet

Foods and Dietaries

A Manual of Clinical Dietetics. Third Edition

Robert William Burnet

Foods and Dietaries
A Manual of Clinical Dietetics. Third Edition

ISBN/EAN: 9783744645652

Printed in Europe, USA, Canada, Australia, Japan

Cover: Foto ©berggeist007 / pixelio.de

More available books at **www.hansebooks.com**

FOODS AND DIETARIES:

A Manual of Clinical Dietetics.

BY

R. W. BURNET, M.D., F.R.C.P.,

PHYSICIAN IN ORDINARY TO H.R.H. THE DUKE OF YORK; SENIOR PHYSICIAN TO
THE GREAT NORTHERN CENTRAL HOSPITAL.

Third Edition.

LONDON:

CHARLES GRIFFIN & COMPANY, LIMITED;
EXETER STREET, STRAND.

1896.

PREFACE.

THE form that this little book has taken was suggested by the habit, which for a number of years I have followed, of writing out in detail directions for patients.

In carrying out this plan, I have endeavoured to state briefly at the beginning of each section the *rationale* of the special dietary recommended. To give definiteness to the directions, the hours of taking food and the quantities to be given at each time are stated, as well as the kinds of food most suitable. In many instances there is also added a list of foods and dishes that are unsuitable to the special case.

References are given, where required, to the Recipes for Invalid Cookery, which form the Appendix, and which have all been very carefully selected.

Repetition in the diets will be found to a considerable extent; but this seems to me unavoidable, in view of the fact that, in many instances, what has to be sought for is a simple physiological dietary.

The discussion of food and diet in health is treated exhaustively in several standard works, and I have not touched upon that aspect of the subject, but have confined myself wholly to Clinical Dietetics, with the hope that a small practical manual will be found useful in many cases of illness where a larger work is out of place.

My grateful acknowledgments are due to several friends, and especially to Dr. James Anderson, for valuable suggestions.

R. W. BURNET.

LONDON, 6 UPPER WIMPOLE STREET, W.
December 1890.

PREFACE TO THE SECOND EDITION.

As only fifteen months have elapsed since the first edition of this work was published, I have thought it best to make no material changes in this edition.

Some additions, however, have been made. A short chapter on the dietetic treatment of Influenza has been inserted, and a number of useful Recipes have been added to the Appendix.

R. W. BURNET.

LONDON, 6 UPPER WIMPOLE STREET, W.
April 1892.

CONTENTS.

CHAPTER V.

CHAPTER VI.

CHAPTER VII.

CHAPTER VIII.

CHAPTER IX.

CHAPTER X.

CHAPTER XI.

CHAPTER XII.

CHAPTER XX.

APPENDIX.

FOODS AND DIETARIES:

A MANUAL OF CLINICAL DIETETICS.

—◦◦—

INTRODUCTION.

A KNOWLEDGE of the physiology of digestion lies at the root of sound practical dietetics, and although it does not come within the scope of the present work to enter into a lengthy examination of different foods and their properties, or into a detailed account of the mechanism of digestion, it may be well to advert very briefly to a few points regarding foods and normal digestion.

For the present purpose foods may be grouped as (1) nitrogenous elements (albuminoids, proteids); (2) carbo-hydrates (starches, sugars, &c.); (3) hydro-carbons (fats); (4) salts and water.

The proper apportioning of these different elements, with due regard to the age, circumstances, and surroundings of the individual, constitutes for healthy persons a well-balanced and economical diet. Probably, however, it very seldom happens, even in health, that an exact balance is struck between the wants of the system and the amount of food that is consumed. For example, when the dietary is full and the digestive organs are vigorous, the surplus food assimilated is stored up, and increase of body-weight takes place. Moreover, even in healthy persons some of the food taken is incompletely digested, and is thrown off in the excreta. Again, when the diet is scanty and insufficient for the needs

A

of the system at the time, the reserves have to be drawn upon, and loss of weight is the result. Fluctuations often occur in the same individual, according to the surroundings in which he is at the time placed as regards climate, activity, and other agencies influencing the processes of oxidation. Thus, a diet that is amply sufficient for a sedentary life would not meet the requirements of an active, open-air life. Conversely, the body when called upon to undertake great muscular activity may be able to utilise a diet that in times of less activity would burden it with waste matter and lead to disturbances of health.

Albuminoids are recognised as the chief tissue-forming foods, and they are therefore needed to supply the body waste that is continually taking place to a greater or less extent. By increasing the activity of oxidation, albuminoids produce in the body increased rapidity of tissue change. In circumstances where oxidation is especially rapid, as in vigorous persons undergoing hard muscular exertion in the open air, larger supplies of food, with a greater proportion of nitrogenised elements in the diet, are required than in those who are less actively employed, or in the same persons when leading a more sedentary life. It may, however, very well be questioned whether albuminoids are ever necessary or desirable in the amounts consumed by many persons, and assuredly whenever we are brought face to face with defective elimination of nitrogenous waste, the albuminoid constituents of the dietary should be limited.

In youth, during the period of growth and building-up of the body, when muscular and nutritive activity are greatest, larger supplies of aliment, and especially of the tissue-formers, are digested and well assimilated than in later life when the period of growth has passed, and when bodily activity is not so great.

On the other hand, old people are apt to make the mistake of adhering to a diet consisting more largely of albuminoids than the diminished wants of the system in advanced age require. The result is a condition of discomfort or of distress induced by the inability of their organs to excrete the amount of nitrogenous waste matter that arises from the use of such a dietary.

The nitrogenous principles in foods, although primarily necessary for the maintenance and repair of the tissues of the body, are known to assist also in the production of force and heat. In the latter properties, however, they are secondary to the carbo-hydrates and fats.

Life is not sustained by a non-nitrogenous diet, but the carbo-hydrates and fats when combined with albuminoids are the great force and heat producers. They do not tend to increase tissue change, and any surplus of them left in the body may be stored up against future needs. Failure on the part of the system to assimilate and elaborate its supplies of the carbo-hydrates constitutes a grave defect in nutrition, notably exemplified in the case of diabetes.

The importance of fresh vegetables and fresh meats as bearers of extractives and salines, apart from their actual nutritive value, has been fully established, but is frequently overlooked in ordinary dietaries. If proof of this were wanting, we have it in the fact that evil consequences speedily follow the withdrawal from the dietary of the important vegetable salts. Such consequences reach their climax in attacks of scurvy and allied disorders.

Concerning the condition of the digestive organs themselves, it will suffice here to mention these four essentials of vigorous digestion, namely, a healthy condition of the mucous membrane, a due supply of normal gastric juice, sufficient nervous stimulus and good muscular tone to ensure proper rhythmic movements. Each of these is important in its own place, as separate consideration of them will show; and if one or all of these conditions be impaired or absent, difficulty of digestion will result, in a more or less pronounced form, according to the extent to which the defect exists.

As regards, in the first place, the condition of the healthy gastric mucous membrane, our knowledge has been gained mainly by direct observation. Those rare instances in which it has been possible to observe, through a fistulous opening in the stomach-wall, the actual appearance of the healthy mucous membrane when it is at rest, and also the changes that occur during the process of digestion, have taught much that is of great practical importance in relation to diet, both in health and in the different degrees of disorder and disease. The well-known opportunity of this kind, that was fully taken advantage of by Dr. Beaumont some sixty years ago, occurred in the person of a healthy young Canadian, Alexis St. Martin, who was accidentally wounded by the discharge of a gun. The shot, entering the left side and perforating the stomach, carried away part of its anterior wall. The man passed through a severe and protracted illness, but ultimately regained his health. The opening, however, never closed, and by means of it Dr. Beaumont made his valuable observations, which have since been confirmed and, in some particulars, amplified and corrected by experiments on animals.

The healthy stomach when empty is contracted, and its surface is pale; its vessels small and tortuous. When, however, food passes down the œsophagus and is received into the stomach, the pink, velvety appearance of the mucous membrane is seen to give place to a brighter, slightly darker shade; the vessels dilate and become more full of blood, and the secretion of gastric juice commences. Experiments have further shown that the same results follow upon the introduction of food through an external opening, or upon stimulation by means of a smooth body introduced from without, and gently rubbed against the inner surface of the stomach. In the latter case the effect soon passes off. If, however, the stimulus be carried to such an extent as to cause irritation, effects exactly the opposite of those described above are produced, namely, contraction of the vessels, suppression of the gastric juice, and the secretion in its place of a quantity of mucus. The importance of this physiological fact will appear subsequently on many occasions.

Absorption is an important part of the process of digestion. If the stomach of a healthy animal be ligatured at the pylorus, and food be introduced by the œsophagus, it will be found that a considerable portion of the food is absorbed by the stomach. If the mucous membrane be not healthy and active the transudation goes on very slowly, or does not take place to any appreciable extent.

Now as regards the second condition, namely, a due supply of normal gastric juice. Proteids that have undergone coagulation are very insoluble, even under the action of strong acids; but they are readily acted upon by the gastric juice. It is necessary that the alkalinity of the food as swallowed should be neutralised, and not only neutralised, but converted into an acid state, in order that digestion in the stomach may proceed satisfactorily. It is a well-known fact that if the products of digestion be removed as they are formed, and if the acidity be kept up to the normal standard, a very large amount of food will be digested by a given, and, relatively, very small, quantity of gastric juice. When the acidity of the gastric juice is neutralised its digestive action ceases, and in such circumstances fibrin may remain in it for a long time without being digested. If, however, the gastric juice that has been neutralised be again brought up to the standard acidity, it becomes as active as it was before. When a large quantity of fibrin is placed in a small quantity of gastric juice digestion soon ceases, but if sufficient hydrochloric acid be added, digestion speedily recommences.

Given, then, a healthy mucous membrane, a sufficiency of normal gastric juice, and an absence of any abnormal nervous interference, the comparative digestibility of food is mainly determined by mechanical conditions, which promote or retard the action of the gastric juice upon it. The greater the amount of surface that is presented to the action of the gastric juice, the more quickly will solution take place. Hence tough, doughy substances, that cling in masses, and firm, unbroken fibres, whether of animal or vegetable tissue, longer resist solution in the gastric juice than friable, minutely divided particles. It need hardly, therefore, be pointed out that thorough mastication of food is of the utmost importance, since by that means minute subdivision is secured, and consequently a vastly increased surface-area is presented for the action of the gastric juice. Moreover, in the case of carbo-hydrates, in addition to the mechanical effects of prolonged mastication, their digestion is begun in the mouth, by the action of the saliva converting them into dextrin.

In speaking of the appearance of the healthy gastric mucous membrane, it was implied that the stomach when empty and contracted is at rest. On the introduction of food, however, or when the stomach is stimulated in other ways, certain movements are set up. On the first entrance of food into the stomach there are but feeble indications of movement; hence it is evidently not the fact of the stomach being filled that causes the movements. If that were the cause, they would be greatest at the beginning, whereas observations show that they increase and become more marked as digestion proceeds, till they attain a rhythmic, churning action. The movements, therefore, are clearly an essential part of digestion, and are not due to any mere mechanical cause. By these regular movements fresh portions of the food are brought under the action of the gastric juice.

In health, as has been already pointed out, a dietary is suitable or well-balanced when it is properly adapted to the wants of the individual, and contains a due proportion of nitrogenous elements, of carbo-hydrates, of fats, and of salines, with water. Clinically, however, the definition does not hold good, for here the dietary must be modified according to the diathesis of the patient, the condition of his different organs, and the disease from which he is suffering. For example, when the eliminative processes are hampered, albuminoids must be cut down; and, on the other hand, when the system is unable to deal with the carbo-hydrates, they must be reduced to a minimum, and so on. Again, take, for instance, functional disorders of the digestive organs. In these cases the times of giving food and the intervals between the supplies, the amount given at a time and the mode of preparing the food, are details of almost equal importance with the consideration of the kind of food to be given. In health the regulation three meals a day are found, by the majority of people, to be the most convenient division, but the sick man cannot take a sufficient supply at a time to last for five or six hours, and consequently he must have his smaller supplies at shorter intervals. The length of the interval will depend on the quantity that can be taken at a time, the kind of food that is given, and the rapidity with which absorption is carried on.

Plain, simple cooking should be the rule in health, and much more does this apply to invalid cookery. It cannot be too good, it cannot be too simple, nor for invalids can the food be too daintily served. The surroundings and circumstances under which food is taken have a great deal to do with the comfort of the patient. The fitful, fastidious appetite of the invalid is whetted by the appearance of a little meal daintily served, while the presentation of a large quantity turns him, for the time, against all food. Sick-room cookery is now much better understood than it used to be, but it is to be feared that many nurses and attendants on the sick, not to speak of medical men, hardly yet appreciate fully the help that attention to such details will bring to those under their care.

CHAPTER I.

DISEASES OF THE STOMACH.

GENERAL CONTENTS : Chronic Gastric Catarrh (Chronic Gastritis : Irritative Dyspepsia : Inflammatory Dyspepsia)—Atonic Conditions of the Stomach (Atonic Dyspepsia)—Ulcer of the Stomach—Cancer of the Stomach—Hæmatemesis — Inflammations of the Stomach (Acute and Subacute Gastritis : Acute Gastric Catarrh).

CHRONIC GASTRIC CATARRH (= CHRONIC GASTRITIS : IRRITATIVE DYSPEPSIA : INFLAMMATORY DYSPEPSIA).

THESE are all different names given to varying degrees of a condition, or to several closely allied conditions, that frequently come before us in patients who complain that they are suffering from indigestion.

Symptoms.—On questioning such patients, they tell us that their appetite is variable and uncertain. They complain of acidity, heartburn, and pain after eating; of pyrosis; of sickness at times, coming on soon after meals, and especially after indulging in any rich or indigestible articles of diet; of thirst, and a sense of gnawing, heat, and soreness at the epigastrium; in some cases, also, there is slight epigastric tenderness on pressure. The tongue is red at the tip, furred at the base, or, in cases of long standing, it is bare, often scored and fissured. Sleep is broken and disturbed, and there is usually some loss of flesh and strength. The bowels are constipated, but constipation may, every now and again, give place to an attack of diarrhœa.

Origin and Causes.—The condition indicated by the foregoing brief outline of symptoms may have its origin

8

in gastric congestion and catarrh, resulting from disease of liver, heart, or kidneys. It may be induced by over-indulgence in alcoholic stimulants, and it is especially common in those who are in the habit of taking spirits on an empty stomach. In persons in whom the predisposition exists, frequently recurring or continued irregularities in diet, such as eating things of known indigestibility, hurrying over meals, working close up to and immediately after a full meal, are sufficient to induce an attack. Worry and anxiety are also common causes; and we have not infrequently seen this condition, which is undoubtedly often connected with the gouty state, alternating with outbursts of eczema.

Irritability of the Mucous Membrane. — From what has been said of the symptoms, it is evident that the condition we are here called upon to deal with is one in which there is great irritability of the mucous membrane. In these circumstances, if the food be bulky, so as to distend the stomach; if it be harsh, lumpy, and coarse, so that it irritates this tender surface; if it be highly seasoned with spices and condiments, the hyper-sensitive stomach will turn against it, the vessels will contract, the follicles will refuse their contents, mucus will be poured out; and if that be not sufficient to protect its walls and to allay the increased irritation, vomiting will ensue, and the whole contents of the stomach will speedily be rejected. Even if the stomach does not thus summarily get rid of its burden, the food will remain for a long time undigested, giving rise to flatulence, weight at the epigastrium, and a sense of fatigue and depression in place of strength and refreshment will ensue.

Small Quantities of Food at Short Intervals. — In such circumstances as these, food must be given in very bland, easily digestible forms, in smaller quantities at a time than in health, and at shorter intervals of time. Moreover, the food must be partaken of very slowly. If the symptoms be severe, the rules will be the same as for a case of gastric ulcer or gastric erosion (*see* p. 17), but in less severe cases the following dietary will be suitable :—

Dietary.—In all but comparatively slight cases it will be better for the patient, at first at least, to have his breakfast in bed, as he is sure to feel somewhat tired after the exertion of dressing. If he is sufficiently well to be up and about, . let him take before rising a teacupful of milk, with enough hot water in it to take the chill off, and let breakfast follow very shortly after he is dressed. It may be here repeated that strict injunctions must be given the patient to take all food very slowly by teaspoonfuls, and at a temperature rather cool than hot, but not actually cold. A typical breakfast will consist of a breakfast-cupful of "tops and bottoms," or unsweetened rusks, made as for infants' food, or a cupful of any prepared farinaceous food.

Some patients can go on taking the same things with relish for a considerable time, while others very soon get tired of one thing, and require a change. As much variety as possible should be introduced into the dietary, and, in addition to the alternatives already given, another change may be made by substituting for plain milk, weak cocoa nibs or peptonised cocoa, made with boiling water, with some milk added in the cup.

10.30–11.—A small glass of milk with the chill taken off, and a teaspoonful of malt extract added to it, or a teacupful of chicken tea with a plain biscuit or thickened with some farinaceous material.

1 *o'clock.*—A breakfast-cupful of good beef-tea ("whole beef-tea," *i.e.*, with some of the meat dried, pounded, and mixed with the liquid), or a breakfast-cupful of strong mutton broth with a finger of toast dipped in it.

4 *o'clock.*—A cupful of milk flavoured with tea, or a cupful of cocoa nibs with a slice of thin bread and butter.

6.30–7.—A meal like that at one o'clock; some variety of broth or beef-tea.

9.30–10.—A teacupful of water arrowroot with a dessert-

or a tablespoonful of brandy stirred into it, or a cupful of Mellin's food.

Food at Night.—If the patient sleeps well throughout the night, there is no necessity for him to have food; but if he wakes up, a glass of milk and water, or gruel, or farinaceous food made thin, and kept in a covered jug under a cosey, should be within reach; or some meat-jelly, and a few spoonfuls of the jelly, or a teacupful of the other food, will often help to procure sleep.

Alcohol.—It will be observed that the only admission of alcohol into this dietary is the small quantity permitted at bedtime, and, as a rule, these patients are better without more. Some, however, either from age or from feebleness of constitution, are so weak that it is necessary to allow a small amount during the day as well, and it will usually then be found best to give it in the form of old spirit, in quantities varying from a teaspoonful to a tablespoonful, either in the food at the midday and evening meals or in a claret-glassful of water, sipped just after the food.

Treatment of Severe Cases.—Severe cases of the kind now under consideration will at first require the same treatment as cases of acute gastric catarrh, or of gastric ulcer, and there will be found many gradations between such cases and the slight ones. It will frequently be necessary to begin treatment on the strictest plan, and as the condition of the stomach improves, to advance gradually to a dietary similar to the one just given.

Food in the Convalescent Stages.—As convalescence progresses, a further step will then be to substitute plain *purées* (see *Appendix*) of chicken or game, or *panada* of chicken, or boiled sole, or whiting, for the liquid at the midday and evening meals. Next, a lightly boiled egg, or a poached egg, with dry toast or stale bread and a little butter, and a cup of cocoa, may be given at breakfast. If that be well borne, go on to boiled white fish of the lighter kinds at breakfast, continuing the pounded meat for dinner.

Gradual Return to Ordinary Diet.—By degrees, allow the patient to return to ordinary but very simple food in the solid form. A slice of roast mutton, or the thick part of a mutton-chop, or the breast of a chicken, with bread, some well-boiled spinach, or other tender green vegetable, will come in first at the midday meal—giving boiled sole or whiting at the evening meal. It should be kept in mind that no cases are more liable to relapse than these, and that upon what may seem a very trifling irregularity of diet. The process of change, therefore, must be gradual, and it is well not to run the risk of a relapse for the sake of getting back to ordinary food a little sooner. If such patients are to keep well they must for a considerable time at least, and sometimes altogether, abstain from eating salted and cured meats, preserved or tinned fish, pastry, "made-dishes," raw vegetables, cheese, nuts, sweets, and especially from any form of stimulant upon an empty stomach.

Constipation.—Constipation is often troublesome, especially in earlier stages of treatment, and the irritability of the mucous membrane of the digestive tract prevents its being obviated by the use of brown bread, fruit, and vegetables. Some simple aperient or an occasional enema will be required, but these should not be too soon resorted to nor too often repeated, lest the habit of depending upon artificial means become established.

ATONIC CONDITIONS OF STOMACH: ATONIC DYSPEPSIA.

Symptoms.—Forming a marked contrast to the foregoing cases in many ways, yet not far removed from them, and in some features presenting a family resemblance, are those patients who come before us complaining of a sense of weight and fulness in the epigastrium with distension coming on a short time after eating. Occasionally there is slight pain, sometimes spasms, but as a rule discomfort and a sense of sinking are complained of, rather than any acute pain, and there is no tenderness on pressure—in fact, moderate pressure, for the time being, affords relief. Acidity, pyrosis, and vomiting are usually absent, but there is a good deal of flatulence, and discomfort is temporarily relieved by eructations. The flatulent distension usually extends to the bowels as well, and gives rise to colicky pains in the region of the hepatic or splenic flexure of the colon. These patients are depressed, inclined to take gloomy views of life and of their own affairs; often irritable and wanting in resolution, and, as a rule, they sleep badly. The pulse is feeble, soft, and compressible. The tongue is large, flabby, and pale, slightly coated or clean, and usually indented at the edges by the teeth.

General Want of Tone.—This condition of affairs, indicative of general as well as local want of tone, comes on for the most part gradually. It may follow upon some form of gastritis; it may be brought on more rapidly by some great and sudden change in the habits, as from an active, open-air life to a sedentary one, or it may be hereditary. No one who has given attention to digestive disorders can fail to have been struck by the frequency with which feebleness of digestion is handed down from parent to child, not necessarily in the same form, but often in a greater degree. Whilst in private practice we find that this atonic dyspepsia comes on gradually in those who persistently disregard the ordinary laws of health, by neglect of proper exercise, by hurrying over meals, and by eating and drinking more than their bodily expenditure warrants, —in hospital practice, on the other hand, the most common causes are found to be deficiency of proper food and an undue indulgence in tea.

Appearance of Stomach.—The appearances of the stomach are quite different from those in cases where irritability is the chief symptom. The stomach looks flabby and relaxed; the surface is paler than usual, smoother, more moist, and covered with a layer of mucus.

In the class of cases now under consideration the regulation of diet, as a means of cure, plays a not less important part than in those where, owing to the very sensitive condition of the mucous membrane, its effects are at once more strikingly apparent. Moreover, since this atonic state of the stomach is associated with flabbiness and want of tone of the system, the general condition has to be improved by exercise and by every means that will help to brace up the patient.

Not so much depends upon the kind of food that is taken, provided it is simple, as upon the way in which meals are arranged and the circumstances under which the patient takes his food. If he eats hurriedly, and especially if he sits down to meals hot and exhausted, the inevitable discomfort follows. An alteration, and adaptation to circumstances of the food and hours of eating, will in many cases effect a cure without much change in the diet itself.

Morbid Ideas and Introspection.—A considerable number of these patients are very much inclined to dwell upon their own feelings and symptoms, and to think too much about what food seems to agree with them and what does not. It is better that they should not take their meals alone, and it is of great consequence that their minds should be occupied with some healthy interest which will prevent them from turning in upon themselves.

Bathing and Exercise.—The first direction to them will be to have a morning bath, tepid or cold, according to the age of the patient, the state of his circulation, and the season of the year. The rubbing which follows, and if necessary to produce a good reaction also precedes, the bath is not the least important part. These patients should not attempt to do much before breakfast, though there is no harm in their being in the open air for a short time before the morning meal. Any hard exercise, however, such as tennis or a long walk at that time, will only exhaust and depress. If exercise is to be taken in the morning, it should be preceded by a light early breakfast, say a cup of cocoa and a slice of bread and butter, or a glass of milk and a biscuit. Many patients, especially the more delicate ones, with small reserve stores to draw upon, find that they cannot take early morning exercise without feeling tired, drowsy, headachy, and, as they often express it, "good for nothing" all the day. The reason is, that they go for too long a time without food. They have had nothing to eat for some twelve or fourteen hours, and they are surprised that if they take active exercise before breakfast they are exhausted for the rest of the day. Let such people try the plan of taking a light, early breakfast, after which they will enjoy and benefit by a morning walk, a ride, or a game of lawn-tennis. Here let it be remarked that one of the best, if not the very best, exercise for the "atonic" is a morning ride; it promotes the activity of the liver and obviates the tendency to constipation. If the patient is not accustomed to this exercise, it should be commenced gently, and gradually extended from an hour to an hour and a half, or two hours if time permits; being preceded, of course, by a small, light meal as already indicated, and followed by the bath and breakfast.

Dietary.—Passing on to the details of the diet, the rules should be, simple food, plainly cooked, no large meals, and consequently no long intervals between the meals.

Breakfast.—Boiled sole, whiting, or flounder; or a slice of fat fried bacon, or a lightly-boiled egg. A slice of dry toast with a little butter, or of brown bread (not new) and butter.

Beverage.—One cup of cocoa, or of milk and water, sipped after eating.

Luncheon.—Chicken or game, with bread, and a little tender well-boiled vegetable, such as spinach, vegetable marrow, or young French beans.

Beverage.—Half a tumblerful of water, sipped after eating.

Afternoon Tea.—A cup of cocoa, or of weak tea with milk, and a slice of brown bread and butter.

Dinner.—(Two courses only.)

Fish of the kinds allowed for breakfast, with melted butter.

A plain *entrée*, or game, with a little mashed potato or potato chips.

For sweets and dessert, a plain biscuit will suffice ; *or,*

A slice of any tender meat, such as saddle or loin of mutton, or the thick part of an underdone chop ; a little mashed or grated potato.

A spoonful or two of any plain milk-pudding, to which may be added some stewed fruit.

Beverage.—Half a tumblerful of water, with from one to two tablespoonfuls of spirit.

It will be observed that the quantity of fluid entering into the above diet is very limited, and this restriction in liquids must be impressed upon the patient as an important element in the dietary. The reason for this is apparent when we recollect what the condition of the mucous membrane is, and how that condition interferes with rapid absorption. If a quantity of fluid be taken at meals—for example, soup, or even plain water—digestion is much delayed, fermentative changes take place, and the results are flatulence and discomfort. If, however, these rules regarding liquids be made too strict, the tendency to constipation is favoured, the evacuations becoming scanty and difficult. This may be overcome by giving half a pint of hot water to be sipped at bedtime, and a like quantity of cold water the first thing in the morning. If these means do not suffice to obviate the constipation, and the dietary already contains a fair allowance of brown bread, fresh vegetables, and stewed fruit, massage to the abdomen should be regularly carried out. Failing all other means, a dinner-pill containing aloes, rhubarb, or cascara with aromatics must be given.

Alcohol should be very much restricted, or not given at all.

If used, the best form is old spirit (ʒss.—ʒi) in a little water, sipped after food, or in the hot water at night.

ULCER OF THE STOMACH.

While it is true that the symptoms of ulcer of the stomach vary greatly in intensity and urgency, the slighter cases, equally with the more severe ones, demand careful dietetic treatment.

Rapidly Fatal Cases.—Cases that commence in an insidious way, with obscure symptoms, often go on rapidly to perforation and a fatal termination. It is difficult to say whether in the very speedily fatal cases any treatment could avail to stop the necrotic process and avert a fatal issue; but in other cases we have the satisfaction of knowing that much may be done by regulation of the diet not only to prevent suffering, but in many instances to effect a cure.

Rest Most Important.—Rest is the chief part of the treatment generally adopted in dealing with an ulcer in any other situation, and the same principle applies to the stomach no less than to other parts of the body. Not only must the patient be kept quiet, and warned against fatigue, overexertion, and strain, but in severe cases absolute rest in bed must be enjoined, at least until the urgent symptoms have subsided.

Physiological Rest.—Without such precautions physiological rest for the stomach will be of little avail. The two forms of rest must be united if the treatment is to be successful. Moreover, the patient's supplies of nourishment being somewhat limited, the saving of strength obtained by rest in bed is often very important, and by keeping the stomach itself at rest, as far as that is practicable and possible, the best chance is afforded for the healing of the ulcer. It is not always necessary, indeed it may not always be desirable, to forbid all food by the mouth; but it must be remembered that the introduction of any food, even the

B

blandest, excites the circulation in the stomach and produces an increased supply of blood, thus tending to irritation of the ulcer, besides stimulating the secretion of gastric
juice.

Causes of Pain.—The acid, gastric juice, too, coming in
contact with the tender surface of the ulcer, causes pain,
and often sets up vomiting, which of all things is most
to be avoided. Pain is also kept up and increased by
anything that increases the peristaltic movements of the
stomach; hence the coarser the food and the greater the
quantity, the more the movement and the greater the pain,
which in such circumstances usually lasts until the offending
mass is rejected.

Coarse and Bulky Food to be Avoided.—All rough and
bulky food must therefore be forbidden, and the patient
must be restricted both in the kind of food that he takes
and in the quantity taken at a time. The food must be
of the mildest kind—pulp and pap, or fluid given in small
amounts at short intervals, and the smaller the quantity that
can be borne at a time, the shorter must necessarily be the
intervals between the supplies.

**Necessity for Food being taken very Slowly and in Small
Quantities: Alkalies.**—Another important point is the way
in which food is taken. The sensitive stomach is much better
able to retain food with comfort if it be taken very slowly
than if it be hurriedly gulped down. In writing out
directions for such cases it is best to insist upon all food
being taken by teaspoonfuls, and that, for example, instead
of attempting to take a draught of even simple cold water,
it should be slowly sipped, or small pieces of ice should be
sucked. When there is no dislike to milk it should form
a main part of the dietary, as milk and soda water, or
peptonised milk, or in the form of bread or rusks and
milk. Where acidity is troublesome, lime water, prepared
chalk, or magnesia is a good counteractive, and when added
to the milk, prevents its curdling into hard masses, which

so often set up vomiting. Beef-tea or broth thickened with some farinaceous substance, such as powdered biscuits, baked flour, or a prepared farinaceous food, may be given two or three times a day. Malt extracts are very valuable additions to the dietary, and can be used either alone or in milk. Sometimes skimmed milk, though, of course, not so nutritious, or even butter milk, agrees best.

Nutrient Enemata. — When, however, a sufficiency of nourishment, even with the greatest economy of bodily waste, cannot be taken by the mouth in the way indicated above, without too greatly curtailing the periods of rest, nutrient enemata must be resorted to in addition. In the severe forms, where the symptoms are urgent—for example, during the occurrence of hæmatemesis—all food by the mouth must be stopped and reliance placed for the support of life upon nutrient enemata alone. With due precautions as to warmth and rest, not only may life be sustained in this way, but even the weight of the body maintained for a considerable time. Treatment is rendered much more difficult in those cases where nutrient enemata are not well borne; but seeing the importance there is for keeping up the patient's strength, they should not lightly be given up, and every means should be tried to secure their toleration. In some extreme cases inunction with oils has been found useful in tiding over an anxious time and aiding other means of support.

Small Doses of Sedative before Food.—In connection with the dietetic treatment of gastric ulcer, it should be mentioned that not infrequently where great irritability of the stomach exists and there is difficulty in getting food retained, a very few minims of laudanum or of liquor morphiæ in a teaspoonful of water, given shortly before the food, prevents vomiting. As has already been remarked, vomiting is of all things most to be avoided. It tries the patient sorely and greatly interferes with the healing of the ulcer.

Treatment of Constipation.—The constipation that is so

commonly present in cases of gastric ulcer cannot be treated dietetically, and is best met by simple enemata.

Importance of an Accurate Record.—It is necessary in most cases of this kind to be precise in giving directions to the nurse in charge of the patient, and the physician will do well to have an accurate record kept of the measured amounts of food given and the exact times at which they are taken. By this means the total quantity taken in twenty-four hours can be at once seen, and the intervals can likewise be noted.

(a.) Dietary in a Severe Case : Nutrient Enemata.—In a severe case the quantity of liquid food given at one time to be sipped by the patient should not exceed from two to three ounces, and the intervals should not be greater than two hours. In this way about two pints of fluid will be taken in the day, and if it be mainly milk with some farinaceous material, such as baked flour, arrowroot, powdered biscuits, or a prepared food like Mellin's or Allen and Hanbury's, it conveys a fair quantity of nutriment, and may be for the time sufficient without nutrient enemata. If, however, the milk, before it will agree with the patient, has to be much diluted with soda-water or lime-water, the nutritive value being thereby so much reduced, nutrient enemata should be given twice in the day, say morning and evening.

Predigested Foods, per Enema. — The best forms of nourishment to be given in this way are prepared and predigested substances, such as peptonised milk, peptonised gruel, beef peptones, malt extracts, and prepared farinaceous foods, with, if necessary, from two teaspoonfuls to one tablespoonful of brandy. It is hardly necessary to point out that, in order to be well retained, the enema must be small in bulk, the temperature lukewarm, and that it must pass gently and slowly into the bowel. A heaped tablespoonful of a prepared malted food, with a little milk and enough hot water to sufficiently raise the temperature, and to make the whole quantity up to one and a half ounces ; or two tea-spoonfuls of beef peptonoids in two or three tablespoonfuls of warm water, are typical examples.

Addition of Opium to Enemata.—If less than from two to three ounces, perhaps only a couple of tablespoonfuls, taken by the mouth, can be borne at a time without causing pain and sickness, then the amount that can be tolerated, even if only from one to two tablespoonfuls, should be given every hour. Three or four nutrient enemata should be given in the day to supplement. Sometimes tenderness and irritation of the rectum and anus prove troublesome, and render this treatment difficult, but usually fifteen drops of laudanum added to the enema, or an occasional opium suppository, remove the difficulty.

(*b.*) **Dietary.**—*In less severe cases,* and, in the absence of urgent symptoms, also during the convalescent stages of the more severe cases, it will suffice to restrict the patient to the blandest articles of diet, food being given in four small meals, with two little intermediate supplies, thus :—

Breakfast, 8 A.M.—A breakfast-cupful of bread and milk, or of rusks and milk, or a cupful of cocoa made with milk, with toast thoroughly soaked in it.

10.30. — A good teacupful of beef-tea or of broth, thickened with some farinaceous material.

1 *o'clock.*—A cupful of plain *purée,* or later on, in convalescence, a piece of boiled sole, with the inside of a slice of bread, neither new nor stale.

4 *o'clock.*—A cup of milk or of peptonised cocoa.

6 *o'clock.*—A meal like the midday one.

9 *o'clock.*—A cupful of milk thickened, or of Mellin's food.

Note.—All the food to be taken very slowly, and neither hot nor cold.

A teaspoonful of malt extract added to the milk twice daily is most useful.

During the night, if awake, the patient should have handy

by his bedside some diluted milk or peptonised gruel, in a covered jug kept warm under a cosey, and of that a cupful may be sipped.

Gradual Enlargement of Diet. — Gradually, as convalescence goes on and the patient gets about a little, the dietary may be enlarged by the addition of sago, tapioca, or maccaroni puddings, custards, eggs beaten up in milk or very lightly boiled, white fish boiled, *panada* of chicken, and good broth ; but still the rule should be a small quantity at a time, and nothing hard, bulky, or difficult of digestion for a considerable time after all symptoms have subsided.

Danger of Relapse. —-When relapses occur, as they not infrequently do, they are almost always traceable to errors in diet. It is important to remember that in gastric ulcer the powers of digestion are but slowly regained, and haste in returning to ordinary diet is almost sure to retard convalescence.

We have *post-mortem* evidence in abundance to show that fresh mischief has often been lighted up either at the site of the original ulcer or in its immediate neighbourhood, and followed by fatal hæmorrhage, simply by errors in diet.

CANCER OF THE STOMACH.

Much that has been said in relation to diet in cases of gastric ulcer applies equally to cancer of the stomach, but in some important points, which have a bearing upon dietetic treatment, the two diseases present different symptoms.

Dilatation of Stomach in Cancer. — Dilatation of the stomach, for example, although occurring in a certain proportion of cases of ulcer from thickening and contraction at the pylorus, is much more common in cancer, owing to the frequency with which the growth causes pyloric stricture. Again, until the pyloric obstruction becomes marked, there is often less acidity present than in ulcer.

Character of Pain.—Pain, likewise, is more constant and continuous in cancer, and it does not bear the same relation to the taking of food as it does in gastric ulcer, being often present when the stomach is empty. It is more of a dull, aching character than in ulcer, but the taking of food may bring on an exacerbation often referred to the back, and resembling a paroxysm of neuralgia.

Loss of Appetite.—What with the loss of appetite, that is so constant a symptom, and the fear of increasing the pain, patients are often afraid to take much food. Others, again, find little difficulty so long as the food is simple and more or less fluid. In these cases, however, violent attacks of pain, lasting sometimes for hours, may occur at night quite independently of the taking of food.

Loss of appetite is usually a marked and early symptom in cancer, but exceptional cases occur when the cardiac orifice is involved, where the trouble is not so much disinclination for food as difficulty in swallowing. Up to a certain point in such cases there may not only not be much loss of appetite, but hunger may be distressing. The food does not reach the stomach, and speedily regurgitates unaltered. Rapid loss of flesh and strength are the result, unless the patient be supported by predigested nutrient enemata, as the only food that can pass into the stomach is a very small quantity of liquid at a time.

Widespread Degeneration of Stomach. — Another important point in regard to cancer of the stomach is, that the tissues of that organ, beyond the limits of the growth, are involved in widespread degenerative changes; hence secretion and absorption are very seriously interfered with, and nutrition thereby so much the more hampered. Progressive, usually rapid, emaciation is a characteristic of the disease, which no dieting can counteract; but the patient may be helped and made more comfortable by the use of nutrient prepared enemata, and the extensive disorganisation of the stomach, already referred to, is an indication for the use of prepared foods and peptonised milk, &c., which will pass through the stomach easily, and be acted upon lower down the intestinal canal. This method of feeding in these cases often gives great relief to the urgent symptoms.

Diet Similar to that in Gastric Ulcer.—For details of diet the reader is referred to p. 17, where, under the heading of *Ulcer of the Stomach*, the particulars are given. The order of things must, however, be reversed in so far that the diet recommended in the later or convalescent period of ulcer is the one suitable to the earliest periods of a case of carcinoma; and since cancer always runs an unfavourable course, the directions which apply to cases of severe ulceration (p. 20) are those to be followed in the later stages of cancer. These are, in brief, liquids in small quantities, at frequent intervals, the length of which will be determined by the amount that can be well borne at one time, and predigested nutrient enemata, given in the manner stated on p. 20.

HÆMATEMESIS.

It is not necessary to give in detail a diet for cases of hæmatemesis, since in the great majority of cases serious hæmorrhage from the stomach is due to either ulcer or cancer of that organ, and in speaking of these affections (pp. 17, 22) the diet during the continuance of hæmatemesis has been mentioned. It is sufficient to repeat here, that during the occurrence of hæmorrhage, and more particularly if it should persist, all food by the mouth should be stopped, and reliance placed upon nutrient prepared enemata to support the patient's strength.

INFLAMMATION OF THE STOMACH (ACUTE AND SUBACUTE GASTRITIS : ACUTE GASTRIC CATARRH).

The heading of this section will of itself indicate the somewhat wide range of conditions that are intended to be in-

cluded in it. It seems best to group these together, considering them, for our present purpose, as different degrees
·of the inflammatory process rather than as different affections.

Pathology.—The *post-mortem* appearances found in the
stomach are often indefinite and misleading, since no organ
is more prone to undergo softening and other changes after
death. Hence these morbid appearances are not all to be
accepted as evidence of pre-existing gastritis. On this
account the observations of Dr. Beaumont on the living subject in the person of the Canadian, St. Martin, are most
valuable as evidences of at least the early inflammatory
changes taking place in the gastric mucous membrane. The
earliest noticeable alterations were reddening of the mucous
membrane, which appeared irregularly congested, with diminished secretion of gastric juice and the presence of a large
quantity of viscid mucus. This state of matters increased;
the patches, besides being more numerous, were larger, and
some of them became aphthous; thick ropy mucus and a
muco-purulent fluid tinged with blood took the place of the
natural secretions. In the instance of St. Martin these appearances gradually subsided under the influence of proper
diet and medicine, but where such a favourable result does
not take place, we can readily imagine that they are the
precursors of more serious lesions, namely, hæmorrhages,
superficial abrasions, and ulcerations.

Gastritis from Irritants.—Cases of poisoning by corrosives
and irritants, where the quantity taken and the intensity of
the inflammation are not sufficient rapidly to destroy life,
are naturally included in this group, as in their after-treat-
ment they require dieting similar to that suitable for other
forms of gastritis. It is likewise manifest that there will
be considerable variety in the intensity of the symptoms in
different cases falling under the above headings.

Considering the complex structure of the stomach, its
functional activity, and the manner in which it is frequently

treated by its owner, acute gastritis, apart from the effect
of direct irritants, is not of so common occurrence as we
might expect.

Predisposition.—Individual predisposition, either inherited
or acquired, is often to be traced, and in such persons inflammatory gastric attacks may be set up by causes that in
others would have no effect. Repeated attacks undoubtedly
render the subject of them more and more liable to a re-
currence of the disorder from slight causes, and any acute
fever or debilitating illness is likewise a predisposing cause
—probably partly because both the quantity and the quality
of the gastric juice are then deficient. If in such conditions of general weakness care be not exercised as to the
quantity and kind of the food taken, the gastric juice not
having sufficient activity, fermentation and decomposition
of food will result, and gastric catarrh will ensue. Thus,
too large a quantity of even digestible food may on this
account act in the same manner as a less amount of indigestible food does. Food, also, that in itself is not indigestible, may become so when imperfectly masticated, in
which condition it is practically impervious to the action of
the gastric juice.

Effects of Alcohol: Chronic Diseases: Chill.—Undiluted
alcohol, a large quantity of condiments and spices, food that
has already begun to decompose before it is taken, and substances that retard digestion, such as narcotics, may each
and all be the exciting cause of acute gastritis or gastric
catarrh. Again, in the later stages of some chronic diseases
—for example, in heart disease and cirrhosis of liver, where
a backtide of blood keeps up congestion of the stomach—
gastritis may be readily induced. A sudden chill in those
who are predisposed to the disorder may of itself bring on
an attack. Over-care in regard to food is considered by
some to be a predisposing cause; but may we not rather
believe that in these persons there is an inherent weakness
of stomach, that experience prompts the carefulness, and

that in reality delicacy is the cause of the susceptibility to acute gastric troubles?

Sthenic and Asthenic Cases.—Coming now to the question of treatment, we may, for practical purposes, divide the patients into two classes—(1) the robust or plethoric individuals, in whom the acute gastric disorder is the result of some irregularity in eating and drinking; (2) those who are weakly and in feeble health, perhaps the subjects of some one of the chronic diseases already alluded to. In the latter, acute or prolonged gastric disturbances are a serious complication.

Fasting.—In the sthenic class abstinence is the first thing to be enforced, both as regards food and drink. Rest is what the stomach needs to regain its normal condition. Even the mildest forms of food cause a certain amount of hyperæmia in the stomach, and so tend to keep up the catarrhal condition. It is, therefore, best to advise the patient to fast for, say, twenty-four hours, more or less, according to circumstances, giving only small pieces of ice to be sucked or sips of iced water. Following upon this will come milk diluted with lime-water or with soda-water, in small quantities at short intervals.

Nutrient Enemata.—If there be any cause for uneasiness as regards keeping up the strength of the patient owing to the fever that may be present, or the persistency of the attack, then nutrient prepared enemata should be given in place of risking more food by the mouth. The question of nutrient enemata is more fully entered into in the section on gastric ulcer (p. 20), to which the reader is referred for details.

Prepared and Predigested Foods.—The same indications for the use of prepared foods hold good as strongly in cases of gastritis as in gastric ulcer, and their use should be continued for some time, until the powers of digestion return more fully. The addition of brandy to the enemata or not will depend upon the symptoms in each individual case, but very frequently from ten to fifteen drops of laudanum are useful in allaying irritability, and securing the retention of the enema.

The prominent point in the treatment, that of giving rest to the stomach, must not be lost sight of; and in proportion as danger to the patient from asthenia becomes apparent, so sustaining measures are called for, and these will consist for the most part of nutrient enemata of the kinds already specified.

Vomiting.—When vomiting is continued and severe, causing great exhaustion, it may sometimes be checked by giving, at frequent intervals, a few drops of brandy in a teaspoonful of iced water, or small quantities of champagne, also iced. In other cases nothing but sedatives, of which the best is morphia, given hypodermically, are of any avail in allaying sickness.

Care in Diet during Convalescence.—During convalescence the greatest caution is necessary in allowing the patient to return to ordinary diet. When fermentation with acidity is troublesome, and cannot be controlled by giving the sulpho-carbolate of soda or glycerine of carbolic acid, it is well to limit the amount of farinaceous substances in the diet, and to give broths, good beef-tea, *panadas*, jellies, &c., in addition to milk with alkalies. Where acidity is not so troublesome, farinaceous materials may be given more freely, and may, indeed, often form the staple of the diet. For further details of quantities see diet (p. 21) for convalescence from gastric ulcer.

CHAPTER II.

DISEASES OF THE INTESTINAL TRACT.

GENERAL CONTENTS: Constipation—Diarrhœa—Acute Enteritis: Acute Gastro-intestinal Catarrh — Inflammation of the Colon — Dysentery : Acute Peritonitis—Typhlitis and Perityphlitis.

CONSTIPATION.

WHILE, undoubtedly, the recognised rule that the bowels should be evacuated once every day ordinarily holds good, there are certainly not a few persons found whose bowels act only once in two or three days, or even at much longer intervals, and whose health nevertheless is perfectly good.

Habit.—It is well to bear this habit or idiosyncrasy in mind when such cases come before us, because where it has become fixed we shall do them little good, and perhaps make them very uncomfortable, by treatment directed towards altering the habit. Even if we succeed in establishing for a time more frequent defæcation, reaction will sooner or later follow, and the bowels will relapse into a state more sluggish than before our interference.

Effects of Diet.—Different habits of diet have a distinct bearing upon the fulness and frequency of the evacuations, and any change in diet that may have occurred should be taken into account in dealing with a particular case. In those who are small eaters, and whose food consists mainly of such articles of diet as are concentrated and leave but little residue, there is, of course, very little waste material to be thrown out by the bowels ; hence constipation is apt to be induced. The indication clearly is to put the patient on

29

a diet more bulky and less nutritious; but in the case of these patients care must be exercised to prevent such an alteration of diet disordering their digestion, especially at first, and the change should be gradually effected.

Oily Matters.—In cases where the fæces are lumpy and the mucous membrane dry, an addition to the diet of some oleaginous materials will of itself often suffice to put matters right. Cream, butter, salad-oil, &c., may be given with advantage.

Neglect and Irregularity.—Where through carelessness and hurry the habit of regular defæcation has been neglected, the necessity of persistent regularity in this important matter must be impressed upon the patient, for without it all our care in arranging a diet will probably be thrown away. In young people this is a frequent cause of constipation. In girls and women, in addition to neglect in this respect, other factors often enter into the causation; for example, the movements of the intestines are often hampered by want of freedom in dress and by insufficient exercise.

Sedentary Habits.—Sedentary habits undoubtedly conduce to inactivity of the bowels, and may lead to chronic constipation where the tendency already exists. A variety of exercise is requisite, and exercise that involves increased action of the abdominal muscles, such as riding, is much more efficacious than merely walking. On the other hand, an undue amount of exercise may induce temporary constipation, and excessive perspiration sometimes has the same effect; but these conditions right themselves, as a rule, without requiring special dietetic or medicinal treatment.

Catarrh.—Chronic intestinal catarrh and other conditions, followed by impaired muscular tone in the intestinal walls, must be recognised as causes of chronic constipation. Loss of general muscular power, and, in women, relaxed conditions of the abdomen following pregnancy, must also be taken into account. In many of these cases systematic massage of the abdomen is invaluable, when conjoined with dietetic means, in relieving constipation and in bringing about a more active state of the bowels. Under its use the tone of the intestinal muscular coat and of the abdominal muscles is greatly improved, the secretions are increased, and regularity of action is gradually established.

Care in Changing the Diet.—In regard to dietetic means in general, it may be said again, care must be exercised in arranging a diet so as not to disorder the digestive organs. Although a certain proportion of innutritious material in the diet is desirable if the patient has been accustomed to take concentrated food, leaving little residue, it will not do to change suddenly to a bulky dietary. The introduction of fruit, green vegetables, porridge, brown bread, &c., must be made by degrees, so that the digestive organs may get accustomed to the change. At the same time, regular and sufficient attention to the calls of nature must be inculcated as being a means of the utmost importance. Articles of diet that disagree and disturb digestion should not be persevered with at first, although perhaps later on they may be gradually introduced and may come to be taken with advantage. In the early stages of treatment it may be necessary also to use some aperient; for example, Cascara sagrada, in small and gradually diminishing doses to aid the diet.

Fruit.—For some people raw ripe fruits do excellently, and can be taken early in the morning or at any time on an empty stomach, which adds to their efficacy; but with others they disagree, and where dyspepsia is a prominent symptom they are contra-indicated. These persons, however, can often digest cooked fruits, especially when prepared with an alkali, as thus the acidity is done away with. Amongst preserved fruits, stewed prunes and figs may be mentioned as decidedly laxative, and so also are raisins. Coffee and cocoa favour the action of the bowels more than tea, and some people find help from a glass of bitter ale or of cider. Many men regard a pipe or even a cigarette after breakfast as conducive to regularity in the action of the bowels, but probably this is at least partly accounted for by the fact that in such circumstances the morning duties are not hurried over, and the bowels get a better chance of being fully relieved.

Hot or Cold Water.—A simple and frequently efficacious aid to regularity of the bowels is to sip slowly a half-pint of cold or of hot water early in the morning. If that does not suffice, the addition of a like quantity taken in the same way before going to bed at night will often produce the desired effects.

In sketching out the details of diet more exactly, we may divide these patients into two classes, namely, those in whom dyspepsia is not a prominent symptom, and those in whom it is markedly present.

Dietary in Simple Constipation.—For simple constipation without dyspepsia the following will be suitable :—

Early in the morning some ripe fruit, such as an orange or an apple.

Breakfast.—Well-boiled oatmeal porridge and milk, fresh fish or fat bacon, brown bread and butter, coffee with milk.

Luncheon.—A slice of cold meat, bread, salad (dressed with oil). Ripe or cooked fruits.

Dinner.—Soup or fish ; any variety of tender meat ; well-boiled green vegetables, and a little potato mashed with cream. Fig-pudding, apple charlotte, or plain stewed fruit with cream. A biscuit and butter. *Dessert.*—Ripe fruit, French plums, or dried figs. A cup of black coffee, without sugar.

Before going to bed, a glass of plain or of aërated water.

Exercise.—A sufficiency of exercise varied in form—walking exercise only often does not answer—should be enjoined in these cases; and if for any reason this cannot be taken, massage of the abdomen should be practised.

Dietary in Constipation Complicated by Dyspepsia. — In the second class of cases, where dyspepsia complicates matters, we have to be very careful not to aggravate this by alterations and enlargements of the ordinary dietary. The early morning fruit sometimes disagrees, raw vegetables are difficult of digestion, and brown bread irritates the sensitive mucous membrane.

A glass of water sipped the first thing in the morning, if taken slowly, will probably cause no discomfort, especially if it be hot instead of cold. If oatmeal is found difficult of digestion, porridge made of whole wheat-meal may be substituted, and coarse brown bread had better be avoided. Salads and cabbage frequently cause flatulence in these patients, and so defeat the objects in view; but young spinach, tender brussels sprouts, and cauliflower can be taken with impunity.

It is obvious that in dealing with this class of cases, we must feel our way gradually, until we learn what the digestive organs in the particular instance before us can accomplish, and we must regulate the diet accordingly. It is safer, and likely to lead to a more speedy cure in the long run, if we begin cautiously, and do not proceed more rapidly than the digestive powers can bear.

Such a dietary would be in detail as follows :—

7.30 A.M.—A half-pint of water sipped very slowly.

Breakfast, 8.30.—A small plate of porridge with a little cream or new milk.

9 *o'clock.*—Some bread, not new, and butter; a small cup of coffee. A few strawberries, grapes, or other ripe fruits in their season.

Or 9 *o'clock.*—If porridge is not taken, a slice of fat bacon or of ham, or a piece of fresh fish, or chicken, or game, should be taken along with the bread and butter and coffee, followed by fruit.

1 *o'clock, Luncheon.*—A slice of any tender meat with

C

bread, and a little well-boiled green vegetable. Beverage, a glass of water.

If meat is taken at breakfast, luncheon may consist of shape or pudding and stewed fruit, with bread or biscuit and butter, followed by ripe fruit.

Tea, 4.30.—A cup of weak tea with milk, and a slice of thin bread and butter, or a cup of cocoa.

7–7.30, *Dinner* (two courses).—Fish or meat and a plain pudding. Beverage, a glass of water.

In the foregoing dietaries, alcohol has not been included, as it cannot be considered a necessary addition, although in a certain proportion of cases a glass of wine at luncheon and at dinner undoubtedly seems to favour regular action of the bowels.

Mellin's Food or Cream.—For invalids troubled with constipation, a breakfast-cupful of Mellin's food is often a sufficient laxative.

In young children a teaspoonful of cream added to one or two of their supplies of milk, or a teacupful of chicken- or mutton-broth, will often obviate a tendency to constipation. Mellin's food in barley-water or dissolved in the broth is likewise a most useful addition in these cases.

DIARRHŒA.

In all cases of continued relaxation of the bowels, treatment by diet occupies an important place, and the dieting of the patient is one of the first questions claiming attention. In fact, where diarrhœa is dependent on local causes, regulation of the diet alone will usually suffice to effect a cure, in the absence, of course, of actual organic lesions.

For the present purpose it is not necessary to enter into a detailed classification of the different forms of diarrhœa and their causes, but one or two distinctions should be emphasised.

(*a.*) **Diarrhœa of Bright's Disease, &c.**—It need hardly be remarked that diarrhœas dependent upon and occurring during the progress of such maladies as chronic Bright's disease, cirrhosis of the liver, tuberculosis, &c., must be looked upon as part of the disease, not as a separate condition, although palliative treatment may be required. These forms of diarrhœa hardly come under consideration by themselves, though some modification of diet, based upon the general principles of treatment of diarrhœa, may be desirable.

(*b.*) **Nervous Diarrhœa.**—Another form of diarrhœa is that set up by causes acting through the nervous system. Persons of unstable nervous system are most subject to this form, and in them such causes as sudden emotion or excitement may induce temporary relaxation of the bowels. If indigestion be present, as it often is in these individuals, the tendency to diarrhœa is much increased, and hence the main object to be kept in view in dieting them is to raise. if possible, the standard of their digestive powers, restricting all such articles of diet as tend to worry a sensitive nervous system. A great deal may be done for the comfort of these patients by putting them upon a very simple diet, easily digestible, and not likely to irritate the mucous membrane of the bowels. This should be particularly attended to whenever they have to undergo any severe mental strain; for example, in the case of students going up for an examination. and in others of a like disposition when subjected to any similar strain.

(*c.*) **Mucous Diarrhœa.**—In cases where much mucus is passed with the motions, we conclude that severe irritation or actual inflammation of the mucous membrane exists, and in these instances much benefit accrues from careful dieting. Farinaceous substances, such as arrowroot, tapioca, or sago, with milk, form a suitable diet, and allay the irritation of the intestinal mucous membrane.

(*d.*) **Fatty Diarrhœa.**—A rare form of diarrhœa is that associated with the passage of large quantities of fatty matter

in the motions. The explanation at first sight would seem to
be an inability on the part of the system to digest fats; but
in some cases, at least, the withdrawal of fatty articles of
diet, and, as far as possible, the withholding of fat-formers,
have not brought about a diminution in the amount of fat
passed. On the contrary, instances of recovery from this
form of diarrhœa are recorded wherein the giving of large
quantities of olive-oil appear to have had a decidedly beneficial effect.

(*e.*) **Temporary Diarrhœa.**—In a passing attack of diarrhœa, due to temporary irritation caused by some indigestible
article of food, the difficulty is usually at an end when the
bowels have fully emptied themselves, and all that is required is to keep the patient quiet and warm, restricting
him for a few days to a very simple diet, with the avoidance
of everything difficult of digestion or likely to set up fresh
irritation. During such an attack the diet should consist of
milk, with some farinaceous food, such as rice, arrowroot,
sago, or tapioca, in small quantities at a time, and at short
intervals. The food should be taken neither hot nor quite
cold, and where there is much prostration, a tablespoonful
of brandy should be given with at least every second supply
of food.

(*f.*) **Chronic Diarrhœa.**—In all cases of chronic diarrhœa
our ultimate object must be to improve digestion and to raise
the tone of the general health. It will commonly be found
that at first, at least, milk and farinaceous foods answer
best, but in some exceptional cases very tenderly dressed
meat and eggs are more easily digested. The rule that
each supply of food should be small, and the interval between
meals shorter than usual, applies here equally as in acute
cases. Broths and strong soups usually tend to relax the
bowels, and are inadmissible for these patients. So also are
vegetables, except perhaps a mealy potato, but ripe fruits in
very small quantities often agree well, and form a pleasant
addition to the dietary. Bael fruit is much used in India

in cases of diarrhœa, and on the Continent acorn coffee is found efficacious.

Alcoholic Stimulants.—A limited amount of alcohol is very often required, owing to the debility that usually co-exists with diarrhœa. It may be given with food in the form either of brandy or of a wine containing tannin, such as claret, burgundy, or St. Raphael.

Raw Meat Treatment.—An entirely opposite diet, namely, that advocated by Trousseau, consisting of pounded raw meat and meat-juice, has been found useful in a class of cases where farinaceous food is not well-borne. The diarrhœa associated with teething in young children some-times yields to this treatment after resisting all other means. The pounded meat may be given with a little salt or sugar, or in jelly, or it may be strained into beef-tea, and given slowly by teaspoonfuls. If this diet does not carry with it sufficient liquid, the quantity may be made up by giving water with some white of egg in it, and nothing else.

Coming now to the details of the different diets.

Dietaries.—(*a.*) *Dietary in Simple Diarrhœa.*—As has already been said, an attack of temporary diarrhœa, due to indigestion, will speedily yield, after the offending substances have been removed, to treatment by diet, aided by rest and warmth.

A teacupful of milk with a little arrowroot or an arrow-root biscuit powdered in it, milk and lime-water, or milk and ground rice with a little cinnamon, may alternate at in-tervals of about three hours, with a like quantity of tapioca or sago and milk. All these should have the chill taken off, but should not be hot, and should be taken very slowly.

(*b.*) *Dietary in Chronic Diarrhœa.*—In all forms of chronic diarrhœa, as previously remarked, the quantity of food given at a time should be small, and no full meals should be taken.

In a severe case of long standing, directions should be given for all the food to be taken very slowly, by teaspoonfuls, and at a very moderate temperature, that is, neither hot nor cold.

8 A.M.—A breakfast-cupful of rusks or "tops and bottoms" in boiled milk.

10.30.—A teacupful of warm milk, with an arrowroot biscuit.

1 P.M.—Any plain milk-pudding, such as sago, tapioca, with an egg in it, or a plain custard.

4 P.M.—A cupful of peptonised cocoa and milk, or of warmed milk, with a biscuit or a rusk.

6.30 P.M.—A meal similar to the midday one.

9.30.—A cupful of arrowroot.

Alternatives will be semolina-porridge, maccaroni-pudding, prepared farinaceous foods, peptonised milk, and beef peptonoids, where these are well-borne.

With the 10.30 A.M. food give, if required, from a dessert- to a tablespoonful of brandy, and from one to two tablespoonfuls at 6.30 or at 1.0 o'clock, and with the last supply of food. In other cases a glass of claret, burgundy, or St. Raphael at 1.0 and at 6.30 will be suitable.

As improvement sets in, a cupful of beef-tea, with plenty of farinaceous material, such as arrowroot or vermicelli, in it, may be substituted for the milk at 10.30 A.M.; next, plain panada of chicken, or a little boiled fish at 1.0 and at 6.30 may be introduced into the dietary.

(c.) *Dietary in a Mild Case.* — For a mild case, or during the convalescence of a severe one, the diet should be somewhat dry, and breakfast had better be taken in bed.

8 A.M.—An egg, lightly boiled or poached, dry toast with a little butter, a small breakfast-cup of cocoa-nibs.

10.30.—A cupful of milk and a biscuit.

1.30.—Chicken or plain panada, bread, and a spoonful or two of milk-pudding. A tablespoonful of brandy in a half tumblerful of water, or a glass of claret or of burgundy.

5 *o'clock.*—A cupful of milk or of cocoa and a small biscuit.

7.30.—A slice of roast or of boiled mutton, or a chop ; a very little grated potato and bread, or dry boiled rice ; a little of any plain milk-pudding or blanc-mange. The same allowance of stimulant as at the midday meal.

10 P.M.—A cupful of arrowroot, with a tablespoonful of brandy stirred in it.

Note.—To each of the preceding diets, if the meals be small, and the total quantity of food taken in the day be not sufficient, a cupful of milk (peptonised) or of prepared farinaceous food, kept warm under a cosey during the night, should be added.

ACUTE ENTERITIS : ULCERATION OF THE SMALL BOWEL : ACUTE GASTRO-INTESTINAL CATARRH.

When the Stomach is not Implicated.—If inflammation of the intestines exists by itself, and the gastric mucous membrane is but slightly or not at all involved, the diet should be selected and arranged so as to throw the main work of digestion upon the stomach. The aim must be to give those things that will be readily absorbed by the stomach, leaving little residue, that thus the intestines may be kept at rest. The foods that most fully answer these requirements are infusions of meat, meat-jellies, and milk. These should all be given diluted, either iced, or at least quite cold, and in very small quantities at a time.

Gastro-Enteritis.—In the great majority of these cases, however, the gastric mucous membrane is affected to a greater or less extent. When this is so, the question of diet is made so much the more difficult. The essential conditions of healthy gastric digestion being then also interfered with, albuminoids are not well-borne, as they decompose, and, by setting up irritation and flatulence, cause additional suffering to the patient. On the other hand, however, care

must be exercised in the use of ordinary farinaceous substances. They, too, are apt to remain in the bowels undigested, inasmuch as the normal secretions are deficient
and inactive. It is in such circumstances that prepared
foods and artificially digested foods come in so usefully.

Prepared and Peptonised Foods: Ice.—What has been said
regarding prepared foods in relation to gastric ulcer and
gastric catarrh holds good in regard to the cases now under
consideration with even greater force. When the secretions
of the stomach and of the intestines are both at fault, it is
of the greatest consequence that the food should be given
in forms that are easily soluble and readily absorbed. Peptonised milk, peptonised gruel (with or without milk in it),
peptonised beef-tea, and soups similarly treated, are all most
valuable in these cases. As already said, small quantities
only must be given at a time, and in most cases it is better
to give the food quite cold, or even iced. Thirst is best
allayed by giving small pieces of ice to be sucked.

Nutrient Enemata.—When absolute rest for the stomach
and bowels is necessary, then nutrient enemata must be
resorted to. Beef-tea, peptonised, and with some prepared
farinaceous food added to it, milk-gruel peptonised, meat
peptones, and preparations of malt, with or without the
addition of brandy, are typical forms of enemata, that convey
a maximum of nutriment with a minimum of bulk. (For
details see section on Gastric Ulcer and section on Prepared
Foods.)

INFLAMMATIONS OF THE COLON: TYPHLITIS: DYSENTERY.

As in inflammations of the small bowel, so here, in
inflammatory conditions of the colon, it is of the greatest
importance that the food should be of such kinds as are
most readily absorbed, and that leave the least residue in
the bowel. This helps to secure rest to the bowel, which is
a step in the direction of cure.

Abstinence from Food.—In recent cases of acute inflammation without previous weakness, it is well to hold back supplies of food altogether for twelve, or even for twenty-four hours. By this means time is given for the removal of retained irritating substances, and, moreover, rest is greatly favoured. If thirst be urgent, it is best allayed by giving small pieces of ice to be slowly sucked, or by allowing sips of iced water, but no large quantities should be permitted. Where it is undesirable to keep the patient so long without any food, milk diluted with iced soda- or seltzer-water may be given in small quantities at a time, or teaspoonfuls of meat-jelly.

The stomach is less seriously implicated, as a rule, in these cases than in those of inflammation of the small bowel, and it is fortunate that it is so, since nutrient enemata are very much contra-indicated.

Concentrated Foods.—During the acute stages the diet should consist entirely of meat-jellies, strong broths, beef-tea with sago, tapioca, or arrowroot; and milk, with saccharated solution of lime, or other form of alkali, in it. It is hardly necessary to repeat that these must all be given in small quantities at a time, and that none of the food should be taken hot.

Liberal Diet in Chronic Dysentery.—In chronic dysentery the diet must be very liberal, because these patients are generally weak and anæmic, but it must also be of the simplest kind, as easy of digestion as possible. Meat of the most tender kinds, such as chicken, game, or mutton, pounded meat, strong soups, strong beef-tea; milk with alkalies, arrowroot, tapioca, and other farinaceous substances, also the prepared farinaceous foods; peptonised beef-tea and peptonised milk-gruel; eggs beaten up in milk, &c. If bread be given, it should be in the form of toast or rusk, well soaked in milk or cocoa.

Alcohol.—Alcoholic stimulants are, as a rule, required, and the form will vary with the circumstances of the case, but first on the list will stand brandy, next dry sherry and good old port. Brandy or sherry may be combined with milk and egg beaten up in the form of egg-flip, or the spirit may be added to the prepared farinaceous foods.

ACUTE PERITONITIS.

In almost every case of acute peritonitis, the question of how to maintain the patient's strength by sufficient nourishment enters largely into considerations of treatment.

Necessity for Rest to the Bowels.—It is absolutely necessary that the intestines be kept as much at rest as possible. The nourishment, therefore, must be in concentrated form, so as not to distend the digestive organs, and of such kind as to leave little or no residue, thus avoiding as far as possible any increase of peristaltic action.

Avoidance of Vomiting.—The difficulty most frequently is that food cannot be retained by the stomach, even when given in the simplest forms and in very small quantities at a time. Vomiting is by all means to be avoided, if it be possible, and therefore feeding by the mouth has often to be greatly restricted, or altogether abandoned. It has already been pointed out. as being most essential, that the bulk of the food should be the least possible, and so, where the stomach does not reject them, small quantities of strong beef-tea, or of beef-jelly iced, or of iced milk, may be given at frequent intervals—a spoonful at a time.

Artificially Digested Foods.—Artificially digested foods are very valuable in such circumstances, and greatly help to fulfil the indications of dietetic treatment in these cases. Peptonised milk is not palatable, but peptonised milk-gruel is much more readily taken, and is highly nutritious (Recipe 76). Soups, beef-tea, jellies, and shapes can be prepared in the same way, and form useful varieties (Recipes 77, 78). Two to three ounces given every two or three hours even, help very much to tide over the worst times, and if the case proceeds favourably, the amount may be gradually increased.

Alcohol.—Alcohol is frequently required in cases of peritonitis. The safest form is old brandy, but later on champagne may be given with safety and benefit. Thirst is best relieved by small pieces of ice allowed to melt slowly in the mouth.

Nutrient Enemata.—In cases where the stomach rejects all food, feeding by the mouth should not be persisted in, as it only increases the distress of vomiting, and adds to the danger of the patient. In such cases, small nutrient enemata must be relied upon to keep up the patient's strength, and the use of prepared and peptonised or predigested materials greatly increases their nutritive value.

Prepared Nutrient Enemata. — Carbo-hydrates, in the form of prepared farinaceous foods, or plain baked flour, with malt extract, peptonised milk-gruel, peptonised beef-tea, &c., with or without brandy as may be necessary, answer admirably for these enemata (*cf.* p. 20). A tablespoonful of such a food as, for example, Mellin's in an ounce of warm water, or two ounces of prepared milk-gruel, or a teaspoonful of beef peptones in an ounce of warm water, thrown carefully into the bowel three or four times a day, does much to sustain life.

(For fuller details regarding nutrient enemata, see chap. " Nutrient Enemata "; also Recipe 79.)

TYPHLITIS AND PERITYPHLITIS.

Rest to the Bowels and Small Residual.—It is of great consequence in cases of inflammation in and around the cæcum that the diet should be carefully arranged and selected, so that small quantities are given at short intervals, and that while the food contains sufficient nourishment, it is small in bulk, and leaves but little residue to be passed down the bowel. The importance of rest to the bowels has been pointed out in speaking of peritonitis, and it is of almost equal importance in the cases now under consideration. The food should leave no bulky or irritating residual to worry the inflamed bowel. As in cases of peritonitis, the indications are best met by giving concentrated nourishment in small quantities at a time, and by the use of artificial digestive agents.

Four ounces of peptonised milk, or of milk-gruel, beef-tea, soup or jelly similarly treated, may be given every two hours. If a somewhat larger quantity can be well-borne by the patient, six ounces may be given every three hours—the gruel, beef-tea, or soup being given once to three supplies of milk. (For the preparation of peptonised foods, see Recipes 74–79.)

Alcohol.—In young and strong subjects alcohol is not required, but in those of feeble and debilitated constitution, as in second attacks, a tablespoonful of brandy may be given with every second supply of food.

CHAPTER III.

DISEASES OF THE LIVER, &c.

GENERAL CONTENTS : Functional Derangements—Acute Congestion of the Liver—Inflammation of the Liver—Diet in Gravel and Uric Acid—Calculus—Oxaluria.

FUNCTIONAL DERANGEMENTS OF THE LIVER.

Symptoms.—Patients who complain of headache, disinclination for any physical or mental exertion, sleeplessness, and depression of spirits, with dyspeptic symptoms, such as furred tongue, disagreeable taste in the mouth, loss of appetite, flatulent distension and constipation, attribute their troubles to the fact that their liver is out of order. In the main, the conclusion is a right one, and in confirmation of this view we find that the urine is dark, often loaded with lithates, and containing bile pigments, while the motions are scanty and of a pale straw-colour, or even clay-coloured and offensive.

Our knowledge of the action and uses of bile leads us to expect that such a combination of symptoms would result in cases where its secretion is disordered, for we know that bile has antiseptic properties, preventing decomposition in the contents of the bowel; that it gives colour to the stools; that it promotes peristaltic action; and that, in conjunction with the pancreatic juice, it has considerable power as an emulsifying agent and solvent of fats.

Mode of Life.—When not associated with organic disease of the liver itself, or of any other important organ, or with malaria, functional derangements of the liver can usually be

45

traced to an unhealthy and inactive mode of life. The patients are generally those whose primary digestion has been good. They have been accustomed to eat and drink freely, and they have taken very insufficient exercise. Continued indulgence in more food than the needs of the body require will of itself lead to derangement of the liver, and when associated with indolent habits or forced inactivity, the disorder much more speedily results, more especially if to these causes be added worry and mental depression.

Association with Gout.—Persons of the gouty diathesis are very subject to liver disorders, which sometimes alternate with attacks of acute gout, and both may be traced to the same cause or combination of causes. The only hope that such people have of keeping well is by leading an active life, and being abstemious, both as regards meat and drink. When a quantity of nitrogenous matter in excess of the wants of the system and beyond the powers of the liver is consumed, part of the surplus is not converted into urea, but remains to be eliminated by the kidneys, as uric acid and urates, or laid up as gouty deposits.

Biliary and Renal Calculi.—The formation of biliary calculi, too, is associated with or dependent upon derangements in the functions of the liver. Usually continued liver disorder precedes the appearance of these calculi, and the same, moreover, may be said regarding renal calculi, of the uric acid kind at least. Persons, then, who are subject to hepatic derangements should be extremely simple in their diet and active in their habits. They should walk or ride on horseback regularly, and should as far as possible avoid spending their lives in hot climates, or keeping their houses and offices over-heated.

Dietaries.—*Articles of Diet to be Avoided.*—Coming to the details of the diet, it may be said that such persons should avoid all rich and highly-seasoned dishes, as curries, pies, and pastry of all sorts, strong soups, foods rich in fats; salmon, herrings, eels, mackerel, and other fish of an oily

nature ; elaborate *entrées*, and made dishes of all kinds ; also
rich sweets and creams; cheese, dried fruits and nuts, likewise malt liquors, sweet and generous wines, such as sweet
champagne, madeira, brown sherry, and port, must be prohibited.

Articles to be Used Sparingly.—They must use sparingly
red meats, eggs, butter, sugar, cakes, puddings, milk and
cream. They must be very moderate in their use of even
light wines. If alcohol is taken at all, the safest form for
many such patients is weak brandy or whisky and water.

The diet may be arranged as follows :—

Breakfast.—A small plate of very well-boiled oatmeal or
hominy porridge and milk, followed by a cup of weak tea
and a slice of toast with very little butter.

Or a bit of fresh white fish or a slice of bacon with the
tea and toast as above.

In summer fresh ripe fruits, such as strawberries or
gooseberries, and other fruits or plain salads in their season
may be added.

Luncheon.—Some vegetable soup with bread ; or fish, with
a little mashed potato or fresh green vegetable, bread and
butter, with lettuce or watercress.

Beverage.—A glass of claret in water, or one tablespoonful of spirit in a small tumblerful of water, or a glass of
plain or of aërated water.

They may be allowed a cup of tea in the afternoon, with
a slice of thin bread and butter, a rusk, or a plain biscuit.

Dinner.—(Two courses only, soup or fish, and meat, or
meat and pudding.)

Plain soup, especially of the vegetable kinds, or fish, as
sole, whiting, plaice, turbot.

Chicken, game, or tender mutton ; mashed or grated
potato and well-boiled green vegetables, such as spinach
or stewed celery.

Any sort of plain milk-pudding, or shape and stewed fruit.

Beverage as at luncheon.

Nothing after dinner but a glass of cold or of hot water, to be sipped before going to bed.

ACUTE CONGESTION OF THE LIVER.

Acute congestion of the liver has to be treated dietetically, in a very different way from the congestions that result from long-standing heart or lung disease. When the liver is acutely congested, besides the symptoms of disturbance in the hepatic functions already alluded to in speaking of functional derangements of that organ, there is enlargement, and usually some tenderness on pressure in the right hypochondrium. There is often also troublesome vomiting, owing to the irritable state of the stomach. So long as the symptoms are acute, the diet must be restricted, and the patient limited to the most simple forms of nourishment, such as milk diluted with an alkaline water, broths, beef-tea, weak tea, barley, or toast-water. If thirst be very troublesome, these latter may be sipped freely, or the patient may be allowed to suck small pieces of ice. The milk, broths, and beef-tea should be given in regulated quantities of a cupful every three hours, and to each supply should be added a teaspoonful of powdered biscuit or other farinaceous substance, such as arrowroot, sago, or tapioca.

Avoidance of Alcohol.—Except in cases where weakness renders it an absolute necessity, alcohol in every form should be withheld.

When, after rest in bed, with warm applications, and the use of cholagogues, and a diet as above sketched, the acute symptoms have passed off, the greatest care is still necessary in regard to what food the patient takes. The diet must be continued of the simplest kind, and for a long time the rules laid down in the section on functional disorders of the liver should be strictly adhered to. Most persons who have suffered from acute hepatic congestion do well to abstain thereafter from alcohol in any form.

INFLAMMATIONS OF THE LIVER (ACUTE AND CHRONIC HEPA-
TITIS): MALIGNANT DISEASE AND MORBID GROWTHS IN
THE LIVER.

Acute Stage.—In the acute stages the diet should be
limited to liquid foods in the simplest forms, as directed
under the head of "Congestion of the liver" (p. 48).

In the early stages of chronic inflammations and of
malignant disease, the patient's condition will be amelio-
rated by attention to the matter of diet. The directions
already given for the dietetic treatment of functional disorders of the liver should be followed, with this caution,
however, that the great need there is in such cases
for maintaining the patient's strength should never be
forgotten ; and hence, while diet cannot be too simple, it
should be as nutritious and sustaining as the impaired
functional activity of the digestive organs will permit.

Alcohol.—In inflammations of the liver, the use of alcohol
is contra-indicated unless the patient's strength be greatly
reduced, as, for example, after the opening of an abscess.
In malignant diseases, on the other hand, wine and other
forms of alcohol may safely be given in such quantities as
the circumstances of each case render necessary at the time.

DIET IN GRAVEL AND URIC ACID CALCULUS.

Uric Acid.—Excess of uric acid in the blood, and deposits
of it in the urine, being very closely allied to the gouty
state, it is not necessary to repeat all that is said regarding the dietetic treatment of that condition (see chap. on
Gout), or of the treatment of disorders of the liver (p. 45),
with which both are intimately associated. It may, however, here again be remarked that the diet must be restricted
in quantity as well as in certain articles. The nitrogenous
elements, especially red meats, must be used in great
moderation ; and all rich heavy dishes should be altogether
excluded, among these being numbered pies, pastry, sweets,

D

sugar, and fruits containing much sugar. Alcohol, if permitted at all—and in some cases where it distinctly aids the digestive processes a small quantity is beneficial—should be given in diluted forms and free from admixture. Therefore, liqueurs, port, sherry, sweet champagne, and malt liquors are to be prohibited. Fish, white meats, with plenty of well-cooked green vegetables, and salads where they agree, should, along with a moderate amount of milk and of farinaceous substances, constitute the diet. Water is the best solvent and diuretic, but the quantity taken at meals should be restricted. Taken in moderation, between meals, and also the first thing in the morning and the last thing at night, it will not interfere with digestion.

In Children.—In the case of children and young persons who inherit the diathesis, a great deal may be done to prevent its fuller development in later life, by attention to these dietetic rules, combined with a simple active life, exercise in the open air, and suitable clothing.

Oxaluria.

Oxaluria may occur during the course of some chronic diseases, as, for instance, in phthisis, or it may appear temporarily in some persons after they have eaten such things as rhubarb, tomatoes, or unripe fruits. If persistent, it is associated with or dependent upon impaired digestion and imperfect assimilation of food.

No distinctive rules for diet, applicable to all cases of oxaluria, can be laid down, but in general it may be said that food should be moderate in quantity, and should not contain much nitrogenous material. The great point is to secure, as far as it is possible, in these persons complete assimilation, and thus prevent accumulation of waste matter. The diet that is most fully digested answers best, and the activity of the digestive organs should be promoted by fresh air and open-air exercise.

CHAPTER IV.

DIABETES MELLITUS.

IN no other disease, probably, are the beneficial effects of a well-chosen dietary so marked as in cases of diabetes; but in considering questions of diet in this connection, we require to keep clearly before our minds three very different conditions, namely—(1) diabetes mellitus; (2) simple glycosuria; (3) diabetes insipidus, so called, for which a better designation is polyuria.

Distinctions.—These three affections, although they touch each other in one or more points, are in reality widely separated as regards pathology, prognosis, and treatment.

Glycosuria not Diabetes.—The more or less continuous passing of traces of sugar in the urine, or the occasional and temporary passing of sugar in the urine in considerable amount, does not constitute true diabetes ; and hence we must not diet all glycosurics as if they were diabetics, else, in some cases at least, we shall do more harm than good.

Diabetes insipidus likewise is mentioned in this connection only to be separated and made distinct from both the others, as, except the passing of a large quantity of urine, it has nothing in common with either.

Definition.—While there are many variations in the clinical history of cases of true diabetes, and all the symptoms need not be present in the same case at one time.

besides the special symptoms referable to the urine, we
shall always find at least some of those that constitute a
well-marked group.   The usually keen, or even voracious,
appetite of the diabetic is, nevertheless, accompanied by progressive loss of flesh, although a large quantity of food is
consumed by the patient.   There must then be some defect
in assimilation.   It is this—a loss of power on the part
of the organism to deal with starches and other carbo-
hydrates.   Our knowledge of the transformations that the
carbo-hydrates undergo in the body during the progress
of digestion is by no means complete, but we know that
they pass through certain stages which bring them into
the circulation in the form of grape-sugar, and that as such
they enter the liver, there to undergo further changes in
the process of assimilation.   It is when these latter changes
do not take place that we have an excess of sugar in the
general circulation, a condition of affairs associated with
other symptoms of the group, namely, a tormenting thirst,
a harsh dry skin, a red beefy tongue, and a sweetish odour
of the breath.   The thirst, which is often a most distressing
symptom, leads to the consumption of large quantities of
fluids, and thus a high blood-pressure is kept up, rather
favouring the rapid flow of saccharine blood from the liver,
and also the secretion of a large quantity of urine.   Urine
containing an increased amount of urea, as well as a considerable amount of sugar, is necessarily of a high density,
and often possesses acid irritating characters.

Pathology.—While it must be admitted that the pathology of diabetes mellitus is still very obscure, there are
certain facts bearing on its dietetic treatment that should
here be summarised, as they tend to clear our way in reference to the diets to be prescribed in different cases.

Mention has already been made of the fact that the power
of assimilating starches and other members of the carbo-
hydrate group is, in cases of diabetes, defective.   The non-
assimilation of these substances. so important to the nutri-

tion of the body, results in wasting, often advancing to extreme emaciation, which is usually accompanied by a trying sense of weakness and weariness. There is, as already said, an excess of sugar in the blood, and it is passed off by the urine in place of being used or stored up for the needs of the body.

Importance of Maintaining Nutrition.—The condition of the patient is serious in proportion as the body is starved by excessive loss of sugar, and we may, in fact, to a great extent gauge the progress of the disease and the success or failure of dietetic means by the gain or loss in body weight that the patient manifests during the treatment. So long as nutrition is well maintained there is good hope that the patient will weather the storm; and, in observing the varieties in the clinical history of this disease in different subjects, we find that the cases that run the most favourable course are those in which, while the chief symptoms remain in abeyance, the weight of the body keeps up or increases. In young subjects who are passing a large quantity of sugar, and who speedily become very thin and weak, there is far less hope of benefit accruing from a carefully regulated diet than in stout elderly patients whose digestive powers remain fairly vigorous, although they may continue to pass a large quantity of highly saccharine urine.

How far a Rigid Diet should be Enforced. — In cases, too, where the cutting off of all starchy and saccharine articles of diet is not, in a reasonable time, followed by distinct amelioration of the patient's condition, the outlook is not hopeful, and it becomes a serious question how far a strict diet should be rigidly enforced. If the adherence to a rigid dietary is followed by loss of appetite and disturbance of the digestive functions so great as seriously to interfere with nutrition, it will be necessary to try very carefully the effect of slight relaxations in the least important points, always, however, keeping up a strict watch on the effect of such changes in the dietary. The patient's reason must be convinced by a sufficient explanation of the necessity that exists for his self-denial, and thus his hearty and loyal co-operation will be secured. Even with the best resolutions on his part, there is danger of the diet becoming very tiresome, if not altogether unbearable, and he will require considerable force of will if he is to adhere strictly to rules.

Combination and Preparation of Foods.—Success in treatment will depend very largely on the trouble taken on our part in the combination and preparation of the foods that are allowable, so as to introduce as much variety into the diet as the necessary limitations will allow. The dietary must be satisfactory, not only with respect to the allowability of its constituents, but it must also, as far as is possible, satisfy the likings and appetite of the patient, and it must be suitable to his digestive powers. The directions must be written with precision; general instructions will not do in these cases. The permissible articles of diet must be clearly stated, and likewise those that cannot be allowed.

Rationale of Diet.—Starches and sugars must be excluded as far as it is possible to do so. By the exclusion of starches and saccharine articles of diet, the excess of sugar in the blood, on the one hand, is reduced, and thus the urgent symptoms are relieved; while on the other hand, by supplying such foods as are highly nutritive, but yield the least amount of sugar, the body is supported. Thus, in view of the possibility of the cause which set up the diabetes being removable, both the negative and positive sides of the question are of the greatest importance.

The statement as to the total exclusion of starches and sugar is qualified for several reasons. Very few patients are found able to go on for many months, or even for years, upon an absolutely rigid diet; they cannot endure it, and we often find that by allowing them a little ordinary bread no serious harm is done. In fact, the strain is relieved, and they are able to carry out the treatment in other details with greater strictness. As said before, however, the effect of any deviation whatsoever from rules must be carefully watched.

Preliminary Steps in Treatment.— Before commencing treatment, the general condition of the patient should be carefully recorded, in order that there may be a definite standard of comparison, and his exact weight should also be taken. The urine too should be examined several times before the special diet is commenced. Then gradually the forbidden articles must be cut out of the dietary, sugar being one of the first to go. The quantity of bread taken must by degrees be limited, and what is taken must be thoroughly toasted, so as to prepare the way for bran or gluten bread. Potatoes are often greatly missed, but they too must go, and so must all foods containing much starch, such as sago, tapioca, maccaroni, &c., until we bring the patient on to the diet recognised as being most suitable for diabetics.

Dietaries.—Before sketching out the different meals, it may be well to give here a general list of things that the diabetic may eat, of those that he may eat but very sparingly, and of those that are injurious to him.

He may eat:—Beef, mutton, and all kinds of butcher's meat except liver. Bacon and other smoked, salted or preserved meats. Game, poultry, and other birds. Broths and beef-tea, or soups not thickened with any farinaceous substance. Fresh fish of all kinds, also salted, kippered, or otherwise preserved fish. Eggs prepared in any form. Butter and cream. Cream-cheese. Cheese, especially the poorer sorts. Bran or gluten bread and almond biscuits. Spinach, Scotch kale, turnip-tops, lettuce, cucumber, watercress, mustard and cress, and all kinds of green salad, endive, radishes, mushrooms. Pickles, vinegar, and oil. Unsweetened jellies, unsweetened creams. Custards without sugar. Walnuts, almonds, filberts, brazil-nuts and coconuts, olives.

He may drink :—Tea, coffee, and cocoa made from the
nibs. Claret, burgundy, chablis, hock, dry sherry, whisky,
soda-water, apollinaris, &c.

He may eat, but very sparingly, of :—Turnips, cabbage,
brussels sprouts, French beans, cauliflower, broccoli, asparagus, seakale, vegetable marrow.

In this list in the instances we have already given will
come ordinary bread or biscuits. In the matter of drinks,
milk, brandy, and bitter ale may be here included.

He must not eat :—Any form of sugar, wheaten bread, or
biscuits. Sago, arrowroot, barley, rice, tapioca, maccaroni,
vermicelli. Potatoes, carrots, beet-roots, parsnips, peas,
Spanish onions. Puddings and pastry of all kinds. Fresh
and preserved fruits in every form.

He must not drink :—Sweet wines, effervescing wines,
port wine, and sherry. Liqueurs, stout, porter, and all
sweet ales, also cider.

Dietary for a Diabetic. — *Breakfast.* — Bacon or cold
meat, preserved meats or tongue; eggs; fresh fish with
melted butter, or fish preserved in oil; gluten bread and
butter. Salad or watercress. Tea with cream, coffee or
cocoa nibs with cream.

Luncheon.—Soup (not thickened). Mutton chop, cold
meat, chicken or game. Potted salmon or trout, lobster
salad or plain salad. Bran or gluten bread. Cream-cheese
or skim-milk cheese. A glass of claret or hock in effervescing water.

Afternoon tea.—A cup of tea with cream. Almond
biscuits.

Dinner.—Soup or broth (not thickened). Fresh white
fish with melted butter, or salmon with lobster-sauce.
Brill with anchovy-sauce, or cod with shrimp-sauce, or trout
with parsley-butter (without flour). Sweet-bread, pig's
cheek boiled, salmon or lobster *mayonnaise.* Pigeons with
stewed mushrooms, ox palate. A plain joint, or mutton
chop, or steak. Spinach, stewed cucumber, or brussels

sprouts. Game of any sort. Gluten or bran bread and butter. Cream. Salad of any kind and cheese. Almond biscuit. Claret, hock, or whisky with any aërated water.

N.B.—For those patients who find it difficult to adhere entirely to unsweetened foods, saccharin comes in as a ready and safe sweetening agent, and is described by Dr. Pavy as an acquisition in the dietary of the diabetic. Owing to its immense strength as compared with sugar (nearly 300 times), a very small quantity only is required, and used with ordinary care in very moderate quantities, no harm results from its employment.

SIMPLE GLYCOSURIA.

The term glycosuria is still not uncommonly used as a synonym for true diabetes, though its use as such is of more than doubtful propriety, seeing how different the two conditions really are, and how much confusion of thought exists in regard to their relations the one to the other.

Distinction from Diabetes.—In speaking of diabetes mellitus, the distinction between that disease and mere glycosuria has been mentioned, and our object now is to endeavour to make that distinction more emphatic.

The passing of a small quantity of sugar in the urine by persons who are apparently healthy is not uncommon, and it has even been asserted that a minute trace of sugar is a normal constituent of healthy urine. This matter, however, has been made a subject of careful inquiry by competent observers, and it seems now to have been conclusively proved that no sugar is to be found in perfectly normal urine. It is certain, however, on the other hand, that small quantities of sugar may be passed for considerable periods of time without being associated with any marked deterioration of health, and without the presence of any of the other symptoms that are found in true diabetes.

Temporary Glycosuria.—(*a.*) In a considerable number of cases glycosuria is a temporary condition, existing during some passing disorder of health, and disappearing when that derangement has passed away. Again, sugar may be found in the urine of otherwise healthy persons after a meal at which sugar and starches have been consumed in larger quantity than the digestive organs of such persons can at the time cope with.

Gouty Glycosurics.—(*b.*) The subjects of simple glycosuria are usually stoutish, middle-aged or elderly people, frequently with markedly gouty tendencies, and more or less dyspeptic. They indulge freely in sweets and all forms of saccharine foods, and, as Dr. George Johnson in one of his lectures graphically puts it, "these are cases in which the taste for such food has survived the power of digesting it." As has already been said, this is a condition entirely different from diabetes, and patients who are the subjects of it do not require to be dieted with the same strictness that is necessary for diabetics.

In the majority of instances the patients are gouty, and besides the presence of sugar, there is often a trace of albumen in the urine. This fact should at once put us on our guard, for we must not forget the existence of such a thing as "gouty kidney," and the greatest care should be taken not to diet the patient in such a way as to help on this tendency.

Danger of Kidney Disease.—If carbo-hydrates are to a great extent cut off and a large supply of nitrogenous food substituted, the extra strain thereby thrown on the kidneys may, in the circumstances, suffice for the production of that form of Bright's disease which we know is not uncommon in gouty subjects, namely, the contracted kidney. If we can, we must avoid the substitution of a greater evil for a lesser. In constructing a dietary, therefore, for glycosurics, this must be borne in mind, and, while the amount of saccharine food is carefully limited, care must be taken also not to burden them with an excess of albuminoid material. Their excretory capabilities are not of the best, and are therefore liable to break down if too heavily taxed.

The glycosuric must not eat:—Sugar, sweets in any form, chocolate, &c.; pastry, rich puddings, tinned fruits, dried or preserved fruits and nuts.

He must not drink:—Sweet wines, sweet ales, stout, porter, liqueurs, sweet cider.

He may eat:—Poultry and game, fish of all sorts, but preferably the lighter kinds, such as sole, whiting, turbot, brill, cod, trout. Sweet-breads, pig's cheek, or calf's head. Brown or white bread toasted. All kinds of green vegetables and potatoes.

He may drink:—Claret, hock, chablis, dry sherry, brandy, or whisky in water (plain or aërated), cocoa nibs or cocoatina, coffee, and tea, except where dyspeptic conditions forbid it.

The following articles of diet should be used in great moderation by all persons of a glycosuric tendency associated with a gouty diathesis :—

Red meats, cured or salted meats, such as ham, tongue, &c. Preserved or kippered fish. Eggs. Fruits of all kinds, both raw and cooked. Cheese; beetroot, tomatoes, and rhubarb; pickles. Bitter ales, burgundy, port wine, brown sherry, madeira, and champagne.

Dyspepsia.—As a rule, gouty dyspepsia is an outstanding symptom in these cases; the patients complaining, some most of flatulent distension, and some most of acidity. The reader is referred for full details to the section on the diet for dyspeptics, and it need here only be remarked that each case must be specially investigated, and individual peculiarities or difficulties of digestion observed, so that the most appropriate diet for each may be arrived at.

POLYURIA (DIABETES INSIPIDUS).

When the diagnosis of simple polyuria has been arrived at, to the exclusion of true diabetes and diseases of the

kidney, the dietetic treatment of the case, in the absence
of any apparent definite cause, resolves itself into the question of how best to support the strength of the patient,
and compensate for the loss the body sustains through the
continuous drain upon it.

Liquids not to be Limited.—Experience proves that limitation of the quantity of fluid consumed by the patient is not
followed by beneficial results, but rather, in these cases,
tends to increase the difficulty of keeping up nutrition.
While, however, the total quantity of fluid taken in the
twenty-four hours must not be restricted, great benefit to
digestion will often be brought about by preventing the
patient from taking much liquid with, or just after, meals ;
and allowing him to use freely at other times simple drinks,
such as barley-water, toast-water, lemon-juice and water,
apple-rice water, tamarind-water, &c. (Recipes 19, 20,
21, 24.)

If digestive disturbance is a prominent symptom, the
diet must be regulated to suit the powers of the patient's
digestion, and in all cases the aim must be to enable
the patient to assimilate an abundance of plain substantial
food, supplemented, if need be, by cod-liver oil and malt
extracts.

CHAPTER V.

DISEASES OF THE KIDNEYS.

ACUTE BRIGHT'S DISEASE.

Removal of Inflammatory Products from the Kidney.— In dieting a patient suffering from acute Bright's disease, one fact to be kept prominently in mind is the difficulty that the system has in getting rid of its waste material, and especially of its nitrogenous waste. Another important point is, that the diet should be such as to assist in carrying off those inflammatory products by which we know that the tubules of the kidney are to a greater or less extent blocked.

In view of the hampered excretion of effete matter, the amount of nitrogenous elements in the food must be kept down, and in view of the second point, namely, to aid in washing out the products of inflammation from the uriniferous tubules, diluents must be given freely.

Water as a Diuretic.—Water is the best diuretic. Hot water and hot diluent drinks are doubly useful, as they not only keep up diuresis without irritating the kidney, but also because they promote the action of the skin, and in this twofold manner increase the excretion of waste.

Milk the Staple Food.—Diluted milk is the food that answers best, and skimmed milk has a high reputation as a diuretic. Butter-milk, whey, and koumiss are also useful. From two to three pints of milk, well diluted, given in the twenty-four hours, will in most cases be sufficient at first ;

61

but if the disease be protracted and tends to become chronic, a more liberal allowance of liquid food must be ordered, and broths may have to be added to the dietary. When milk alone is used, it should be given in divided quantities at stated intervals—half-a-pint every three or four hours, diluted with half as much hot water or effervescing water.

Diluents.—In the intervals between the supplies of milk, the patient may sip freely of diluent drinks, such as barleywater, toast-water, or imperial (Recipe 21), but care should be exercised that they be not taken too hurriedly, or in such quantity at a time as to upset the stomach.

CHRONIC BRIGHT'S DISEASE.

Before considering the distinctions between different kinds of chronic Bright's disease, it is desirable to notice some points that, in relation to diet, are common to all forms of chronic renal disease or renal inadequacy.

Excretion of Nitrogenous Waste.—If we bear in mind the fact that the kidneys are the great agents in the work of excreting nitrogenous waste, it need hardly be pointed out that if these organs are unable fully to perform their functions, an accumulation of nitrogenous waste products must necessarily take place in the blood. This is so to a greater or less degree in proportion to the extent to which structural changes have advanced in the "large white," and in the "lardaceous," as well as in the "cirrhotic" kidney. As soon, therefore, as we have evidence sufficient to prove that the kidneys are labouring, and are burdened by their work, we must endeavour to remove the strain by regulating the diet; and one clear indication is to limit the supplies of nitrogenous food.

Influence of Alcohol.—Another important point is the influence of alcohol on these patients. In cases of chronic renal disease, alcohol is badly borne, and unless there is some urgent reason for its use, it should not be employed. While, therefore, such important hygienic measures as the wearing of warm clothing and other means of protection from cold, together with regular habits of living in every respect, are not to be forgotten, great importance attaches to the regulation of the patient's dietary, especially in regard to the use of animal food and of alcohol.

Origin of Chronic Renal Disease.—There may be a history of an attack of acute Bright's disease at some former time, which has left the kidneys permanently and progressively damaged, and has rendered them unable to stand the strain that, through ignorance of ordinary physiological laws, has been habitually imposed upon them. In many cases we can trace an habitual over-indulgence in food and drink. In some, chronic dyspepsia is prominent; and if we go a step farther back, in other cases we shall find that the liver is primarily at fault.

If the condition has not had its origin in an acute attack, it has probably crept upon the patient stealthily. The kidneys have for long been struggling to eliminate an excess of nitrogenised waste matter, and they have broken down in their efforts to rid the system of these products of imperfect digestion.

The discovery of albuminuria is made, perhaps, quite un-expectedly, and it is to be feared that too often the treat-ment is based rather upon an endeavour to control the loss and to supply the place of the albumen lost, than to relieve the kidneys from the strain under which they are suffering.

Danger of Over-Feeding.—In other words, the tendency is to over-feed the patients, and to forget that, though the loss of albumen in this way may be a serious matter, the real danger to the patient is the progressive degeneration of the kidneys, and the increasing retention in the blood of deleterious waste that necessarily ensues.

We may very probably not be able to remove the worry and mental strain that play an important part in the clinical history of renal cases, but we can at least so select the diet as to make the work of the liver, and consequently also of the kidney, as light as possible.

Necessity for Relieving the Kidneys from Over-Work.—Excesses in eating and drinking may or may not be traceable as the cause of the renal disease in the particular case with which we are at the time dealing, but the consumption of a large amount of animal food and the free use of alcohol undoubtedly must be stopped if the mischief in the kidney is to be arrested. As has been already said, the supplies of nitrogenous food must be cut down, and alcohol must in most cases be struck out of the dietary altogether. We say in most, not in all cases, because there probably is here and there an isolated example where the digestive processes do not go on so smoothly and perfectly without alcohol as when it is allowed, but the amount must be small, and the effect must be carefully watched.

When to give Alcohol.—If we can assure ourselves that digestion is more completely carried out with the aid of a small quantity of alcohol, then it must not be withheld. We must, however, be very sure, and be ever on the watch both as to the quantity taken and the effect on the patient's general condition, as well as upon the amount of albumen passed.

Fluids and Diuretics.—Another matter of importance is the question of the quantity and kind of the fluids taken. It may be said in general that simple drinks, such as plain water, toast-water, barley-water, or the good old-fashioned cream of tartar and lemon drinks, are all useful as diluents, and that they probably aid the action of diuretics. This caution, however, must be given—they should be sipped slowly, and not gulped down in large quantities, and they should be taken in the main between, and not at meal-times, else they will frequently interfere with digestion, and do harm instead of good. In cases of granular kidney, especially where the changes in the vessels throughout the body are advanced, care is necessary in the amount of liquid taken, and in the manner in which it is taken, lest by over-pressure hæmorrhages be induced or increased. In other forms of renal degeneration, too, where the heart is seriously implicated and there is much dropsy, if we find that liberal allowances of fluid have not a diuretic effect, and are not helping to free the system of retained poisonous matters, it is time to ask ourselves whether they are not doing actual harm.

Milk.—Milk has a great reputation in cases where the kidneys are at fault, but it must be remembered that when it enters into a mixed dietary, milk does not always agree well, and hampers the digestion of other foods, in some cases even causing a copious deposit of uric acid to appear in the patient's urine. In bad cases the use of milk as the sole article of diet is often attended with the happiest results, and this is so more particularly in the case of children. A purely milk diet generally suits young subjects, and its employment for a time is followed by abatement of the symptoms, and by the disappearance of albumen from the urine.

In a certain proportion of cases of chronic Bright's disease, persisting or recurring dyspepsia is a troublesome and difficult symptom to deal with, more particularly where there is also marked anæmia. In such circumstances questions of diet become more complicated, and each individual case must be judged by itself, the dietetic treatment that seems best at the time being adopted. It may, however, be here remarked that these anæmic patients are not always benefited by a fuller dietary with an increased supply of animal food.

Milk Diet Alone.—If the digestive organs are so enfeebled that they cannot utilise the supplies of food given, no good, but only harm, will result from relaxation of the rules already laid down, and such patients will often be found to do best upon a diet of milk alone. It is remarkable how long this diet exclusively of milk can be maintained in the

E

case of those patients with whom it agrees. Instances are well known of patients who have adhered to it for years with marked benefit to health, and while leading fairly active lives. Sometimes skimmed-milk only can be borne, as the cream disagrees and causes dyspepsia, or the patients become too stout. On the other hand, the removal of the cream renders milk more constipating. Where, therefore, unskimmed-milk agrees, and the patients do not put on too much flesh, it forms an admirable diet, supplying all that is wanted for the bodily needs, without excess of nitrogenous elements, and affording sufficient fluid to wash away effete materials.

Relapses.—In those cases where exacerbations of the symptoms occur from time to time, milk diet should be had recourse to, and the duration of that treatment will depend upon the circumstances of the case and the urgency of the symptoms. In the intervals a varied but carefully selected dietary should be employed. As a general rule, the milk will agree best if warmed, and taken in quantities of from half a pint to a pint at a time. The total amount consumed by an adult in the twenty-four hours should be from three to four quarts, if the strength is to be well maintained.

Dietaries.—Putting aside for the present cases in which a milk diet only is required, and coming to details of the ordinary diets, it will be apparent, from what has already been insisted upon, that animal food, such as beef, mutton, veal, lamb, or pork, should hardly find a place in the dietary of these patients.

Game and white meats, such as chicken, may be allowed in moderate quantities, say, once a day.

Fish is not only allowable, but is a useful form of albuminoid. Fat bacon and ham need not be excluded if they can be well digested. Broths, with rice or barley and vegetables in them, fish-soups, and soups made with milk or cream, are all useful articles of diet. Well-cooked vegetables and salads, where digestion is good, may be freely allowed, with the exception of beans, peas, lentils, &c., which have too much nitrogenous matter. Fruits, cooked and raw, are admissible. Pastry had better be omitted, but bread and butter, cakes not too rich in eggs, puddings, to which the same remark applies, and biscuits are not objectionable.

The dietaries, giving variations for different days, might be sketched as follows :—

For the First Day.—*Breakfast.*—A plate of oatmeal, whole wheaten meal, or hominy porridge, with cream or good milk ; bread or toast and butter ; cocoa, tea, or coffee, with plenty of milk added.

Luncheon.—A bit of fish, with a little melted butter, some mashed potato and green vegetable. Biscuit or bread and butter.

In the afternoon.—A cup of tea with milk, a slice of thin bread and butter.

Dinner. — Soup, *purée* of potato, chicken, or rabbit, mashed potato, green vegetables, plain or milk pudding, with stewed fruit. *Dessert.*—Ripe fruit.

Beverage.—A glass of aërated water.

The last thing at night.—A glass of milk and soda- or seltzer-water.

Second Day.—*Breakfast.*—A slice of well-mixed bacon or fat ham may take the place of porridge, and be followed by toast, by bread and butter, tea, coffee, or cocoa, with plenty of milk.

Luncheon.—A basin of vegetable soup, a bit of cheese, bread and butter, and salad.

Afternoon tea.—Toast or rusk, tea filled up with milk.

Dinner.—A piece of boiled fish, butter-sauce, a plain *entrée* with vegetables, milk-pudding or shape, stewed fruit or blanc-mange, biscuit or bread and butter, a glass of plain or aërated water.

At night.—Milk and soda-water.

Third Day.—*Breakfast.*—A bit of fish, fried or grilled, toast or bread and butter, coffee, cocoa, or tea, with plenty of milk.

Luncheon.—A milk-pudding with stewed fruit and cream, bread and butter.

Afternoon tea.—Cocoa or milk or weak tea, a slice of bread and butter.

Dinner.—Fish-soup, game or poultry, mashed potato, green vegetables, maccaroni-cheese.   *Dessert.*—Ripe fruit.

Beverage.—A glass of plain or aërated water.

At night.—A glass of peptonised milk or of milk and soda-water.

CHAPTER VI.

SCURVY.

Curative Effects of Diet.—Experience is not wanting to prove that cases of scurvy are, in the absence of complications, rapidly cured by dietetic means, and that without a proper arrangement of the diet, medicinal and hygienic measures, although valuable aids, will not of themselves effect a cure.

Scurvy is nowadays so much less frequently seen, that the possibility of its existence in a mild form, under the influence of a very restricted diet, coupled with bad hygienic surroundings, is apt to be forgotten, and it is very probable that cases in which the unhealthy blood-state exists to a partial extent, and without the marked symptoms of the disease, are sometimes overlooked.

The exact nature of the changes in the blood has not yet been fully made out, but we know that scurvy never occurs except in circumstances where the diet is restricted, wanting in variety, and deficient in fresh fruits and green vegetables. When fresh vegetables and fruits cannot be obtained, lemon-juice is found to be the best preventive as well as a most efficient remedy.

Usually the digestive powers are not much impaired, and the patient is able to take ordinary food freely when it is reduced to pulp, as the inflamed state of the gums prevents the mastication of solids.

Danger of Syncope.—Bearing in mind the fact that in severe cases there is a great tendency to syncope, which has often proved fatal, the diet should at first be so arranged that the patient may obtain an abundance of nourishment, and still retain the recumbent posture even when taking his food.

Necessity of Soft Food.—Milk, broths, beef-tea, eggs beaten up in milk, potatoes mashed with milk, green vegetables, such as spinach mashed, pounded meat, chicken *panada*, ripe fruits, will be suitable early in the treatment, and a pleasant beverage will consist of lemon-juice in water slightly sweetened.

(*a*.) **Dietary.**—*For a severe case*, the following directions will be suitable :—

7.30 A.M.—A cupful of milk, warmed.

8.30.—An egg beaten up in milk, or a cup of cocoa, with thin bread and butter, or a plate of well-boiled porridge with cream. These may alternate on different days, and should be followed by some fresh ripe fruit.

11.30.—A cupful of good broth, with plenty of fresh vegetables in it.

2 P.M.—*Panada* of chicken pounded and warmed, with bread crumbs, mashed potatoes, well-boiled and mashed green vegetables. Ripe fruits or stewed fruits and cream.

5 P.M.—A cup of cocoa with rusks or toast soaked in it.

7.30.—A meal like the two o'clock one.

10 P.M.—A cup of milk or prepared food.

Lemon-juice and stimulants.—As a beverage, lemon-juice, as already mentioned, will be grateful in most cases.

Malt liquors seem to have a decidedly favourable effect in most cases. In some, burgundy or brandy and water agrees best.

(*b*.) *In less severe cases*, patients will be able to take larger meals at a time, and consequently the intervals between times will be longer.

The general directions as to the selection of foods will

apply equally in these cases as in the severe ones, and the divisions will be into four meals. A good breakfast, a midday meal, afternoon tea, and dinner. In the lesser degrees of the malady stimulants are not so necessary as in the severe cases.

CHAPTER VII.

ANÆMIA.

THOUGH our knowledge of the life-history of the blood is by no means complete, we are conversant with conditions arising in connection with well-marked changes in its constitution, and that are associated with defects in the processes of its formation.

Causes of Anæmia.—We know that the condition which we call anæmia, wherein the blood is deficient in its most important constituents, may arise under very different circumstances. It may occur in the course of such diseases as syphilis or chronic Bright's disease, and it may be associated with lead- or mercury-poisoning, or with malaria. In all these cases the anæmic condition is the outcome of the continued action of the poison in the blood, and the aim of treatment must be to remove the primary evil before much good can be expected from dealing with the anæmia.

Simple or idiopathic anæmia, however, is a totally different condition from any of the foregoing, and the diagnosis can generally be clearly made out; leukæmia also being excluded.

The Anæmic Type.—Simple anæmia, of which alone we have now to speak, comes before us usually in the person of a young girl, about or just beyond the age of puberty, pale, puffy, and breathless, with impaired digestion, feeble circulation, and usually persistent constipation. For our present purpose, we must consider first the state of the digestive organs, and we find that dyspepsia is more or less

72

prominent. Obviously the diet must be modified according
to the patient's powers of digestion, the constipation, that, as
already has been said, is an almost invariable symptom,
being also dealt with before the good effects of dietetic
treatment can be obtained.

Simple Diet and Rest in Bed.—The more marked and
long continued the anæmia, and the greater the consequent
weakness, the more simple must the diet be at the commencement of the treatment. My experience of severe
cases of anæmia has led me to the conclusion that the
speediest way to a cure is to keep the patient in bed for a
week or even for two or three weeks, if the symptoms are
very pronounced and the constitutional disturbance great.
These patients have gradually failed because their expenditure
has been greater than could be met by lessened powers of
assimilation, and the rest in bed, especially in the case of
girls who for some time have been struggling with their
work, does wonders in bringing back the power of the
digestive organs and restoring the balance between the
wants of the system and the assimilation of nutriment.

Hygiene.—On the other hand, I believe that in many cases,
in those, at least, in which the condition is of but recent
origin, the patients would soon get well without any special
treatment if we could place them in favourable hygienic circumstances as regards good food and fresh air, securing also
the full and regular action of the bowels.

Use of Iron.—The known beneficial effects, however, of
iron, combined with aperients in the treatment of anæmia,
lead us to resort speedily to its use in most cases without
waiting for the result of other means alone. Thus, perhaps,
we are inclined to give to iron too large a place in the
treatment of anæmia, to the neglect of such other valuable
means as a regulated diet, the use of aperients, fresh air,
and suitable exercise, by all of which better assimilation
and improved nutrition are brought about and established.

One point in this connection should be emphasised,

and it is this, there is little or no good to be expected from the exhibition of iron when the digestive organs are seriously out of order and the bowels constipated. Therefore, it is often well to wait a short time before commencing to give iron in any form; and should this disordered state of things recur during the treatment, the iron had better be omitted for a short time, and then resumed gradually.

Severe Cases. — In severe cases of anæmia, then, the patient should be kept in bed for a short time at the commencement of treatment in an airy room, with, if possible, a sunny exposure, and fed upon the simplest and most nutritious articles of diet in such forms as the weakened digestive organs can best utilise, and at the same time with as much variety in the diet as possible. At this stage, meat cooked in ordinary ways cannot be digested, but strong broths and soups with pounded meat in them, and thickened with some farinaceous substance, milk and soda-water, milk-pudding, and other dishes made with milk, well-boiled green vegetables, and cooked fruits, form the staple articles of diet.

Prepared and Artificially Digested Foods. — In cases where there is extreme difficulty of digestion, it is necessary for a time to resort to the prepared farinaceous foods, or the milk, the soups, and other things may be peptonised by the addition of, say, liquor pancreaticus (Recipes 74–78).

Alcohol and Cod-Liver Oil. — Two other things coming fairy under the head of foods, alcohol and cod-liver oil, should be here considered in their relations to these cases. Alcohol is seldom indicated or required in anæmia, but in cases of great feebleness it may be useful in the form of a glass of plain claret or light burgundy with luncheon or with dinner. In other cases, a glass of light brisk ale with these meals may be of service by stimulating the appetite and promoting digestion. Under any circumstances, however, if given at all in anæmia, the quantity of alcoholic stimulant should be small, and it is hardly necessary to add that it should be given with food, not between meals or on an empty stomach. These patients are young persons, and it is highly important that they should not come to be dependent on the continuous use of alcohol. The stimulating effects are not what is wanted, and alcohol is useful in these cases only in so far as it promotes assimilation and improves nutrition.

The other is cod-liver oil, and there are several points that should guide us in its use. If cod-liver oil at all upsets the digestion, it should not be persevered with, although it may be again given a fair trial after the disturbance has subsided, and in many cases it is a useful addition to the food supply, while it increases rather than diminishes the appetite. It can be given along with iron, and in that combination is often of decided help in enriching the blood. Where plain cod-liver oil cannot be taken, it is sometimes well borne in the form of an emulsion, with or without extract of malt, or the extract of malt may be given alone with the food.

Coming now to the details of the diet, and bearing ever in mind the general principle already laid down, that the more digestive and constitutional disturbance there is, the simpler must be the food, we should, in a case of severe anæmia, diet the patient as follows :—

Diet No. 1.—8 A.M.—A breakfast-cupful of bread and milk or rusks and milk, or of some prepared farinaceous food, with a teaspoonful of malt extract. (*Note.*—The bread and milk may be peptonised.)

11 A.M.—A breakfast-cupful of soup or of beef-tea, with pounded meat and some baked flour or biscuit powder in it. Mutton or veal broth, chicken-tea or rabbit-soup, may be used for the sake of variety, and should all be thickened as above directed.

2 P.M.—A milk-pudding with one egg in it. (*Note.*— The digestibility is considerably increased if the dry fari-

naceous material be first exposed to heat in the oven, and some malt flour added to it before the hot milk is poured over it.)

4.30.—A cupful of peptonised milk (Recipe 74), with a slice of thin bread and butter.

7 P.M.—A breakfast-cupful of broth or soup as at 11 o'clock.

9.30.—A milk meal, as at 8 A.M.

During the night one to two cupfuls of warm peptonised milk or of peptonised gruel may be taken.

Gradual Enlargement of Dietary.—As the digestive powers improve and the strength of the patient returns, the dietary should be gradually enlarged, the prepared and peptonised foods being replaced by ordinary articles of diet. Broths, with plenty of vegetables in them, pounded meats, well-boiled porridge, boiled fish, chicken, &c., will gradually lead up to chops and joints, while well-cooked green vegetables, stewed fruits, and plain puddings will by degrees be added.

Diet No. 2.—In cases where it has been necessary on account of the feebleness of the digestive organs to put the patient at first upon the simplest possible dietary, this second scheme of diet may be pursued when convalescence has fairly set in. In less serious cases it will be suitable from the first of the treatment.

Breakfast.—A lightly boiled egg, or a bit of plain fresh fish; brown bread and butter or toast; cocoa and milk.

11 *o'clock.*—A small glass of warm milk, or a cupful of soup or broth.

Early dinner.—A chop or plain cutlet, or a slice of boiled mutton or any other light meat, such as game or poultry; mashed potatoes; well-cooked tender green vegetables; biscuits and butter.

Beverage.—A glass of plain or of aërated water.

Afternoon tea.—A cup of cocoatina or of warmed milk, a slice of thin brown bread and butter.

Supper.—A breakfast-cupful of soup, with the meat pounded and put into it, together with some farinaceous material which has been exposed to heat. A small milk-pudding, made as directed in diet No. 1.

Beverage.—A glass of milk and soda-water.

The last thing at night.—A glass of milk and hot water, or of gruel. (*Note.*—In early stages the gruel may be peptonised.)

CHAPTER VIII.

SCROFULA.

WHILE scrofula is undoubtedly very often inherited, it may, nevertheless, manifest itself in the children of parents who themselves could not be said to be actually scrofulous, though constitutionally delicate, or their health at the time their children were born may have been enfeebled by chronic disease or by age.

Acquired Scrofula.—Among the causes of what may be called acquired scrofula comes first of all improper feeding, usually associated with want of fresh air, and with altogether bad hygienic surroundings. This combination of causes obtains largely amongst the children of the poor in towns. In the case of poor country children, it can hardly be said that fresh air is wanting to them during the day, but their food is often insufficient and wanting in nutritive value.

Insufficient Food.—The poor in many parts of the country have, for instance, great difficulty in obtaining for their children a sufficiency of good milk, and unhappily oaten cakes and home-made bread, both of which contain important elements of tissue formation, have been superseded in too many homes by ordinary baker's bread. The butter, too, that should be liberally supplied, is dear, and its place is too frequently taken by a scraping of jam or of treacle.

Results of Improper Feeding.—Amongst the children of the well-to-do classes, excessive and improper feeding, by keeping up chronic indigestion, acts almost as prejudicially

78

as an insufficient supply of food or a diet of low nutritive value, and a deterioration of health as surely follows.

It is now many years since Sir James Clark, amongst others, pointed out the great importance of a properly arranged diet in reference to the health of children, and emphasised the fact, that even where the element of hereditariness cannot be traced, a disordered condition of system, induced and maintained by chronic digestive difficulties, often leads up to the manifestation of scrofula or of struma. It is apt to be forgotten that a condition of deficient nutrition may be kept up by an excessive and unsuitable dietary, as well as by one that is insufficient and innutritious. The lowered standard of health and strength that ensues is followed by scrofulous manifestations in those who, under more favourable circumstances, might have escaped them.

Wet Nurse *versus* **Unhealthy Mother.**—In the case of the mother being pronouncedly strumous or phthisical, or even markedly delicate, the infant should not be brought up at her breast. A healthy wet-nurse, if possible, should be got, or failing that, the milk of good country-fed cows should be obtained. It will give such children the best chance of growing up strong and healthy if they are reared in the country under favourable hygienic surroundings. After the first few months some farinaceous material will be added to the dietary, and of prepared infants' foods suitable for use in this way there are many excellent varieties.

Dietaries.—By degrees the quantities of food other than milk will be increased, until a diet such as the following will be suitable from about the age of twelve months :—

7 A.M.—A teacupful of milk, with a good teaspoonful of Mellin's food dissolved in it.

10 A.M.—A teaspoonful of fresh whey with a tablespoonful of cream in it.

1 P.M.—The yolk of an egg beaten up in a teacupful of milk, or a teacupful of chicken- or veal-broth, with a teaspoonful of farinaceous food in it.

4 P.M.—A teacupful of milk, or a teacupful of whey with cream in it.

6.30.—A teacupful of beef-tea or mutton-broth with a powdered rusk or other farinaceous material stirred in it, or a teacupful of barley-water with the farinaceous material in it.

During the night.—If the child wakes up, he may be given a little milk, and if he wakes early in the morning he should have a teacupful of milk then, and not be compelled to wait till breakfast-time.

As the child grows older, even if the manifestation of the scrofulous tendency remain in abeyance, the care regarding diet must not be relaxed. The child should be encouraged to take plenty of time over his meals, cultivating the habit of eating slowly, and learning to masticate all foods very thoroughly.

From the age of two or three years the diet might be as follows :—

Breakfast, 8–8.30 A.M.—A basin of bread and milk, or a teaspoonful of cocoatina in a cupful of boiled milk with bread and butter.

11 A.M.—A teacupful of new milk, with a rusk or a plain biscuit.

1.30 P.M.—A small piece of tenderly cooked roast mutton, pounded ; a spoonful of mashed potato with gravy ; a tablespoonful of milk-pudding with stewed fruit, or stewed fruit, and cream in place of pudding.

4.30.—Bread and milk, as at breakfast, or a cup of milk with a spoonful of farinaceous food in it, or milk and cocoa with bread and butter.

Before going to bed, a cup of milk with a biscuit, and the same if he awakes early in the morning.

Use of Cod-Liver Oil.—So much for the diet of young children. Then there comes the question of cod-liver oil, which, since its first introduction, now a good many years ago, has been very extensively used in the treatment of such cases as those we are now considering, and in numberless instances with undoubted benefit. Although some children never overcome their first dislike to it, the majority take it readily, and many seem even to miss it when by chance it is omitted.

Different Types—Cod-liver oil is a convenient form of giving fat, and the fatty constituents of food are much wanted, especially in one type of scrofulous children—the slim and thin, with a strong dash of the nervous diathesis in their composition. In these cases there can be no doubt of the value of cod-liver oil and malt. Better results are obtained under its use than in the thick-set, flabby, and somewhat heavy children, in whom the development of fat is not wanting, and whose nervous systems are by no means unusually active. In these latter, the processes of nutrition are sluggish, and every means of promoting assimilative activity should be employed to induce more healthy growth, and to quicken their sluggish faculties.

Alteratives and Courses of Waters.—It is on this ground, no doubt, that alteratives, such as mercury and iodine, are sometimes useful, as are also courses of iodine and other waters ; the active open-air life followed during a course of treatment at a watering-place is doubtless a most valuable adjunct.

Inunction.—In cases of great feebleness, where cod-liver oil cannot be taken, benefit is often obtained by the inunction of that or some other oil. Whilst cod-liver oil is the most efficacious, its unpleasant odour is a serious objection to its use, and olive-oil will often be found a preferable substitute.

We have discussed the diet of young children who are strumous or have a tendency in that direction, and it remains only to be said that wherever throughout youth the proclivity exists, a nourishing diet should be maintained, and should be aided by plenty of fresh air and exercise in the open air, guarding only against violent games, lest

F

injury should be done to the bones or joints, that are so apt
to give serious trouble in the case of scrofulous children.

Uric Acid Diathesis.—Except in those cases where there
is much deposit of uric acid, as sometimes happens, milk
should be given freely, and indeed should form a staple
article of diet. Well-boiled oatmeal porridge ; brown bread
and butter ; milk-puddings ; cream in different forms ; fruit,
especially cooked, are all useful and desirable. It has been
suggested that where, on account of the uric acid tendency,
fruit would seem to be contra-indicated, it may be given if
potash be used in the cooking, as one would do in the case
of the gouty. There has been a prejudice against bread and
potatoes for these patients, and no doubt a diet composed
largely of these, without more nourishing food to supplement,
is most injurious to organisms requiring liberal supplies of
more concentrated nutriment; but mashed potatoes with
cream or milk along with meat are excellent, and no
objection can be taken to brown bread and butter occupy-
ing a subordinate place in the dietary. For a growing
youth of the type we are now considering, the diet may be
arranged as follows :—

Breakfast, 8.30.—Well-boiled porridge and new milk,
or milk with a spoonful of cream in it. A slice of fat bacon
or an egg with cocoa and brown bread and butter.

Dinner, 1 o'clock.—Meat in any tenderly cooked form ;
mashed potato with milk or cream; well-boiled green
vegetable ; suet-pudding or milk-pudding, or stewed fruit
with cream, or a plain cream and dessert.

Beverage.—A glass of water.

Tea, 6.—Fish with melted butter ; or an egg, or a slice
of fresh tongue ; bread and butter with cocoa.

Supper.—A glass of milk, or a cup of malted farinaceous
food.

If it is relished, and does not spoil the appetite for
breakfast, a glass of warm milk may be given at 7 or 7.30
A.M.

Alcohol.—It will be observed that alcohol is not included in this dietary, and there is seldom any necessity for a breach of the general rule that young people are better without any form of alcoholic stimulant. If the necessity for it is clearly indicated, the form and quantity must be decided by the circumstances of the case.

CHAPTER IX.

GOUT.

CHRONIC AND ACUTE.

A VERY great deal has been written about gout in general, and a great many theories have, at one time and another, been started regarding the pathology of the gouty state; but with these different theories I do not propose here to deal, except in so far as they have a bearing on the dietetic treatment of the disease.

Since gout is usually admitted to be the outcome and manifestation of an undue accumulation of nitrogenous waste in the system, the first questions respecting the dietary of gouty subjects that we have to ask ourselves are, what is the cause of this accumulation? and what are the means by which it is brought about?

Excess of Uric Acid.—We know that the products arising from the oxidation of albuminoid foods in the body are uric acid and urea. In health, when the balance between the amount of albuminoid food taken and the waste, in the form of urea, got rid of, is well maintained, the quantity of uric acid in the blood is extremely small. In gout, however, not only can uric acid in considerable quantity be separated from the blood, but it may be found also under similar circumstances in the fluid of serous effusions. The balance is disturbed, and waste products that should be eliminated are retained in the system.

There are good reasons for attributing to the liver the chief part in the formation of urea, and it seems that so long as the liver can cope with the supplies of nitrogenous food ingested, the surplus waste is carried off by the kidneys as soluble urea, and the system is not burdened by an excess of uric acid.

Hereditariness and Predisposing Causes.—The hereditariness of gout is undeniable in many cases, but the circumstances in which we find it most frequently developed *de novo* are where the individual continuously indulges in a diet composed largely of animal food, takes but little exercise, and is free in the use of wine or beer. In such a case there would seem to be two main factors at work—an excess of nitrogenous food with deficient metamorphosis and elimination. In other cases the amount of albuminoid foods consumed may not be excessive, but the metamorphoses are not carried out to completeness. This is a sufficient cause of gout in many persons, and especially in those who have a marked hereditary predisposition to the disease. The worst combination of circumstances is where the activity of the liver is deficient and elimination defective, and yet the individual, hereditarily predisposed to gout, disregards dietetic rules and leads a life of inactivity.

Lead-Poisoning.—To the above list might be added chronic lead-poisoning, the subjects of which are very liable to develop gout. Fortunately instances embodying such a combination as we have just enumerated are not very common, for though the gouty man may not deny himself good living, he is frequently a person of much energy and considerable activity. Gouty people have often very good appetites, and often, up to a certain point, good digestive powers. It must not, therefore, be forgotten that the mere cutting down of meat in the dietary is not sufficient, if they continue to indulge in rich and undigestible dishes, and take as a whole more food than is necessary for the proper supply of the wants of the system. Although it may be doubted whether continued dyspepsia, *per se*, be sufficient to originate the gouty diathesis, there is abundant experience to prove that dyspepsia, sedentary occupations, and everything that interferes with the functional activity of the different organs of the body, tends to set up the disease, and to keep it up in those in whom the predisposition to it exists. The continued use of acids, for instance, or the taking of foods that tend to the undue production of acids during digestion, has a distinctly retarding effect on the elimination of uric acid, probably owing to the diminution of the alkalinity of the blood. Sedentary habits have already been spoken of, and other causes of imperfect elimination of waste products, such as defective action of the kidneys, inaction of the skin, &c., should be mentioned, for they are all potent in increasing the gouty habit.

Effect of Wines and Malt Liquors.—The free use of wines and malt liquors has undoubtedly a large share in the production of gout and in the keeping up of the gouty state. Port wine has been credited with the most powerful influence in this direction, but champagne, sherry, madeira, burgundy, and all generous wines, have an almost equally marked effect. To most gouty subjects all malt liquors and ciders are almost as bad; but in some cases, where a very small amount of meat is eaten, a little malt liquor seems to be taken with impunity. Many gouty people find the use of milk in any considerable quantity, if persevered in for a short time, almost as trying to them as wine or beer, but there are some with whom, in the absence of butcher's meat, it seems to agree.

As to fruits, some would forbid them altogether to the gouty, but I cannot agree with that view.

Use of Fruits.—*Alkalies added to Fruit.*—I believe that the use of fruits, and especially of cooked fruits, is often beneficial, when they are taken in moderation. Very sweet and indigestible fruits, however, are not allowed, and there are some gouty dyspeptics with whom fruits of all kinds seem to disagree. They produce acidity; but even in the case of these persons the difficulty may often be got over by adding some bicarbonate of potash to the fruit before it is cooked. About an eggspoonful to the pound of ripe fruit, rather more in the case of unripe fruits, is sufficient.

Fats.—Neither do I believe that fats should be excluded from, or too much limited in, the dietary of the gouty, except in those comparatively rare cases in which they disagree and cause acidity. ·Where fats can be taken without upsetting the stomach, they are useful as heat-producers, and they have probably less tendency to increase a deposition of fat in the body than the carbo-hydrates have. Then, again, as to sugar, although uric acid formation cannot be traced back to it, gouty persons should be very careful in their use of sugar, and they should altogether avoid ordinary sweets and sweet foods, on account of the tendency there exists for these materials in the gouty to turn acid. We have already seen that excess of acid is a most undesirable condition of affairs, and indirectly at least favours the production and continuance of gout in the system.

Necessity for Simplicity of Diet.—*Limitation of Nitrogenous Elements.*—In the matter of fruits, however, as in all other· details; each case must be studied by itself. No general rule can be laid down beyond this, that second only, if indeed it should be put second, to the limitation of nitrogenous substances, simplicity of diet is of the first importance, and is important, moreover, in several respects. As has already been remarked, gouty people have often good appetites, and they should not be tempted by a variety of dishes; in other words, their meals should be very simple. Dyspepsia, too, is an enemy ever ready to attack the gouty, and is frequently induced by indulgence in a too rich and complicated diet. While the dietary, then, must be a simple one, it must not be insufficient for nutrition, though the nitrogenous element should be small. Much animal food, therefore, is objectionable, and is so in proportion to its richness in nitrogenous material. White flesh, such as

chicken and rabbit and fish, is preferable to beef, mutton,
and lamb.   Game, probably, stands midway between the
strong red and the white meats.   Sweetbreads, tripe, calf's
head and pig's cheek are also admissible ; but veal, pork,
and all cured meats, with the exception of fat bacon, have
to be omitted from the dietary.   Of fish, the sorts to be
preferred are the whiting, cod, sole, plaice, trout, and turbot.
Salmon, herring, eels, mackerel, and all the coarser fish, such
as pike, hake, and bream, are less desirable.   Soups, and
especially vegetable broths, are not to be prohibited, nor are
shell-fish to those who can digest them without difficulty.
Green vegetables, with a few exceptions, such as asparagus
and tomatoes, are to be freely admitted.

Eggs, and dishes made with eggs, such as puddings, omelettes, and custards and cakes, have to be strictly limited.
The same has to be said of cheese, and in the form of toasted
cheese, souffles, or ramikins, should not appear in the dietary
of the gouty.   Nuts, dried fruits, and preserves had also
better be omitted.

Value of Simple Fluids.—Gouty persons require a considerable amount of fluid to carry off waste products from
the system, and it is to this solvent and eliminative action
that the benefit often obtained in these cases by a course of
"water treatment," either at home or abroad, is to a great
extent to be attributed.   It is no doubt combined with, and
aided by, change of air and scene, more or less active exercise in the open air, early hours, and simple living—all of
which it is difficult for the patient to keep up at home.
Under even his ordinary circumstances, however, the gouty
man may obtain benefit from increasing the quantity of
simple fluid that he consumes, provided he complies with
this caution—not to take it at meal-times or directly after
food.   What is taken with meals should be inconsiderable
in amount, and should be taken towards the close of the
meal.   A large glass of hot or of cold water, sipped the last
thing at night and the first thing in the morning, is often of

immense use by aiding the action of the bowels and liver, and helping to carry away excess of waste material.

Water as a Diuretic.—In an article on the action and use of diuretics published some years ago by Dr. Lauder-Brunton, he extols the value of pure water as a diuretic and as a means of removing waste products. In the case of gouty people he is in the habit, he says, of advising that a large tumblerful of water, with or without some carbonate or nitrate of potash in it, be sipped night and morning. He mentions a case in which a friend of his own, who had been a martyr to gout, kept himself quite well so long as he adhered to a rigid diet of white fish, or of fowl, fat bacon, bread and butter, milk-puddings, green vegetables, and a little claret. If he omits one or two of the glasses of hot water, symptoms of gout speedily show themselves.

My own experience has been that gouty people and persons of constipated habit, in whom also free action of the skin is not easily képt up, and who take little liquid beyond two or three small cups of tea or coffee in the day, are often greatly the better for a morning and evening glass of water taken in the manner already mentioned.

Articles of Diet to be Avoided.—To sum up, then, the following articles of diet must be avoided by all those whose livers are inadequate, and who have a gouty tendency, and it holds good even more strongly in the case of those who are actually the subjects of gout in some of its many manifestations.

Of Meats: Pork and veal; duck and goose; all dried, salted, potted, or preserved meats, with the exception of fat bacon.

Of Fish: Mackerel, pilchard, and eels; all smoked, salted, kippered, or potted fish.

Eggs, sauces, and pickles, vinegar, lemons, and other acids. Tomatoes, asparagus, and rhubarb. Pies and pastry of all kinds, "made dishes," blanc-mange, and all preparations containing much gelatine; rich puddings, omelettes, and cakes containing eggs; skim-milk cheese; nuts, dried fruits, jams, sweets of all sorts.

Beverages to be Avoided are: Malt liquor in every form, port, madeira, champagne, moselle, burgundy, and all other generous wines.

Articles of Diet to be Used Sparingly.—They must be sparing in the use of mutton, lamb, and tender beef (any but the tenderest beef is out of the question); salmon, herring, and mullet, shell-fish; game; cheese, mushrooms, broad beans; sugar; sweet fruits, especially uncooked; milk; dry sherry and spirits.

Dietaries—Details of Diet.—Having stated the negative side of the question, we give now the diet in detail.

The first thing in the morning the patient should sip a large glass of hot or of cold water.

Breakfast.—A small piece of white fish or a slice of fat bacon with little or no lean, a slice of bread and butter or of toast, not buttered hot, and a cup of weak tea with milk.

Luncheon.—A basin of some vegetable broth or of fish-soup, with bread; or a milk-pudding, such as rice and milk, sago and milk, or maccaroni, followed by bread and butter. A glass of water.

Afternoon Tea.—A cup of tea, or of cocoa nibs, with biscuits or toast.

Dinner.—Fowl or rabbit, pigeon, or game, or calf's head. (Mutton or tenderly-dressed beef may be allowed occasionally.) A little mashed or grated potato. Some green vegetables. A little stewed fruit, with milk-pudding or shape.

Beverage.—A couple of glasses of claret, or one to two tablespoonfuls of whisky in half a tumblerful of water.

If it is found more desirable to have the most substantial meal in the middle of the day, the above dietary will answer if the dinner given be substituted for luncheon, which will then serve for supper.

Half a pint of cold or hot water before going to bed.

ACUTE GOUT.

All that has been said in the previous section regarding the diet in gout applies to the dietetic treatment of the gouty diathesis and of gouty patients, between the attacks.

Restricted Diet during Acute Attacks.—During an acute attack the diet must be restricted, but the extent of the restriction will depend upon the age, habits, and condition of the patient. If the subject of the attack be old and of feeble constitution, he will not bear the lowering measures that are not only useful but necessary in the case of a young and full-blooded person.

Meat and Alcohol as a Rule Excluded.—As a rule, it is best in all cases to exclude meat, more especially red meat, from the dietary, and, unless there be some special necessity for its use, to forbid also alcohol. In old people, however, with feeble circulation, accustomed to the regular and free use of alcohol, it will often be necessary to allow a small amount, but the quantity should be carefully regulated. Not more than from two to four tablespoonfuls of old brandy or whisky will in most of these cases be required in the twenty-four hours, and the spirit should be given with or just after food, well diluted in plain or in potash water. If the kidneys be seriously at fault, the use of even this small amount of alcohol is of doubtful benefit, and its effect must be carefully watched.

Sthenic Cases.—In a young and strong subject the diet should consist mainly of farinaceous substances, and broths, made not too strong. In detail the diet may stand as follows :—

8 A.M.—A cupful of bread and milk, or cup of weak tea with milk in it, and a slice of dry toast with a little butter.

10.30 A.M.—A cup of weak vegetable broth, or two ounces of milk with Vichy water or potash-water.

1 *o'clock.*—Rice, sago, semolina, or other farinaceous pudding, made without eggs.

4 o'clock.—A cup of weak tea with milk and toast.

7 o'clock.—Broth or some farinacous pudding as at lunch.

10 P.M.—A cupful of thin gruel or of arrowroot.

Hot Water and Diluents.—He should sip half a pint of hot water twice or thrice daily between meals, and barley-water or toast-water may also be allowed. After the acute symptoms have subsided, boiled fish or a little bit of chicken may be given once a day, and the amount of nourishment otherwise very carefully and gradually increased.

Asthenic Cases.—In old or feeble and debilitated persons, the diets, as already pointed out, although kept on the same lines, must be more supporting, and therefore the broths, &c., given may be stronger, and beef-tea, chicken-tea, or an egg beaten up in milk may be added once a day to the list. Alcohol will also be necessary, under the restrictions already mentioned.

Protracted Cases.—If the attack is a long one, the patient becoming reduced and the pulse feeble, it will be necessary to relax the rules still further, and to allow a somewhat more nutritious diet, including fish, soups, and white meats, as well as the allowance of brandy or whisky.

CHAPTER X.

OBESITY.

MANY persons who are decidedly very stout are yet apparently healthy, and it is a difficult matter to say at what point obesity becomes a morbid condition, though that it does so in a certain proportion of cases cannot be denied.

Whilst a certain plumpness of body is desirable, yet an excess of fat, besides being unsightly, is a burden to its possessor, and many plans for reducing the superfluous fat have at different times been suggested.

Causes.—(*a.*) *Excess of Food.*—Obesity may be due to excessive indulgence in food, but this alone is probably not a common cause, and, moreover, thin persons are often much larger eaters than those who are stout.

(*b.*) *Alcoholic Beverages*, especially malt liquors and sweet wines, have a decidedly fattening effect in the case of many persons, and corpulent people are often large consumers of fluids.

(*c.*) *Sex and Heredity.*—Women more commonly than men become very stout, and heredity is to be looked upon as an important element in the clinical history of a case. Stoutness is to be found prevailing in some families, while others are lean, and cannot be fattened by any means.

(*d.*) *Want of Exercise.*—The want of sufficient exercise, by diminishing tissue change, favours obesity, and the stouter a person grows, the more difficult does the taking of exercise become.

(*e.*) *Gouty Diathesis.*—Excessive fatness is also at times associated with the *gouty diathesis*, and to be effectual in such cases, treatment must be directed against that state.

(*f.*) *Anæmia.*—In anæmia, also, there is often a tendency to unnatural stoutness, which likewise yields to appropriate treatment.

It is rarely found that very stout people live to a good old age, but instances are on record of persons who having been very stout, have been able to reduce their corpulence, and have lived on many years in the enjoyment of excellent health. If any plan or treatment for the reduction of corpulence be adopted, it should be commenced very gradually and carried out with great caution.

Sudden Change Hurtful.—Evil results have often followed sudden changes in diet, especially a great increase in the nitrogenous materials of the dietary, to the exclusion of the carbo-hydrates, by persons desirous of rapidly losing their superfluous fat. No sudden change in the mode of life should be rashly undertaken, and all extreme measures should be avoided.

Fatty Degeneration.—It is not to be forgotten that the heart and blood vessels in these patients frequently undergo fatty change, and also that the kidneys, if not actually diseased, are often quite inadequate to the strain thrown upon them by a diet from which sugar and starches are almost entirely excluded, and which, therefore, consists very largely of nitrogenous materials. Upon such a diet, however, Mr. Banting reduced his weight by over three stones in a year, and, moreover, a return to ordinary diet was not followed by an increase of weight. The chief objection to the plan called after Mr. Banting is that the diet consists too largely of nitrogenous substances, and for many stout people it is dangerous to throw so much extra work upon the liver, and more especially upon the kidneys.

"Banting."—Mr. Banting, in advocating his plan, forbids sugar and starches; milk and butter; salmon, herrings, and eels; veal and pork; potatoes, parsnips, beetroot, turnip, and carrot; bread, except when toasted; pastry and puddings; beer and stout, champagne and port. He limited the quantity of liquids as well as of solids, and in reality a comparison of diets will show that his is not more than sufficient to maintain life in a state of inactivity.

His diet was as follows :—

Breakfast.—Fish, bacon, beef, or mutton; one breakfast cupful of tea or coffee without milk or sugar, and one small hard biscuit, or one ounce of dry toast.

Dinner.—Fresh white fish, beef, mutton, lamb, game or poultry; green vegetables; a slice of dry toast; cooked fruit without sugar; two to three glasses of claret or of sherry.

Tea.—A cup of tea without milk or sugar; a biscuit or a rusk; two to three ounces of cooked fruit.

Supper.—Meat or fish, with toast, and a couple of glasses of claret or of sherry in water.

The amount of meat allowed at breakfast was five to six ounces, a like quantity was permitted at dinner, and about half as much at supper.

	Liquids.	Solids.
Breakfast, ...	6 oz.	9 oz.
Dinner, ...	12 ,,	10 ,,
Tea, ...	4 ,,	9 ,,
Supper, ...	4 ,,	7 ,,
	26 ,,	35 ,,

Preponderance of Proteids.—A glance at this dietary shows the large preponderance of nitrogenous materials over the carbo-hydrates, and also the fact that the diet is a very spare one.

Ebstein's Diet.—Another plan, and one which has the recommendation of more variety, while it is also free from some at least of the objections of that followed by Mr. Banting, has been proposed by Professor Ebstein of Göttingen. He excludes carbo-hydrates, sugar, all sweets, and potatoes. Bread is limited to about three ounces a day.

Besides potatoes, he excludes carrots, turnips, parsnips, and beetroot; but allows asparagus, spinach, cabbage, and the leguminose vegetables freely. No sort of meat is prohibited, and he allows the fat as well as the lean to be eaten.

Professor Ebstein does not set strict limits to the quantity of liquid consumed, and he believes that the rules are better kept in the main, and in the most important points, if some articles of diet which strictly would be forbidden—for example, potatoes—are allowed in very moderate quantities. The diet is thus kept from being so irksome, and the patient is able to persevere longer with it than he could otherwise do. It will be observed that in this plan, while the fats are not excluded, the rule against the carbo-hydrates is very strict.

Gradual Change.—In most cases it will be well not to be too rigid, for the reasons already mentioned, and it cannot be too strongly insisted upon that the change of diet should be made gradually. Sugar, sweets, puddings, pastry, cakes, cream, sweet wines, stout and beer, should all be cut off. All farinaceous articles should be gradually reduced. Brown bread and toast in limited quantities should be substituted for ordinary bread, or its place may be partially taken by gluten bread. Lean meats, including poultry and game, also eggs, should be taken in moderation. Fish may enter largely into the dietary, and all green vegetables, salads, and cooked fruits may also be allowed. As already said, all malt liquors and sweet wines must be excluded. Hock, claret, or other light wine may be allowed to the extent of three or four glasses a day, or a small quantity of spirits in water may be permitted; but the total quantity of liquid taken in the twenty-four hours should not exceed two and a half or three pints.

The corpulent should be encouraged to take exercise, and especially exercise in the open air; but any violent exertion is most undesirable, from the strain it would put on the heart and blood vessels.

CHAPTER XI.

DISEASES OF THE LUNGS AND PLEURÆ.

GENERAL CONTENTS : Pulmonary Phthisis—Asthma—Pulmonary Emphysema
—Chronic Bronchitis—Pleurisy with Effusion.

PULMONARY PHTHISIS.

As in struma, digestive disturbances, with evidences of mal-
assimilation and mal-nutrition, frequently precede the direct
manifestations of the diathesis; so in cases of pulmonary
phthisis, we are accustomed to see the signs of lung-mischief
preceded, often for some considerable time, by chronic dyspepsia, with a disordered state of system and a lowered
standard of health.

Wasting.—Phthisis is essentially a wasting disease, and
the great object of all treatment, dietetic, therapeutic, and
climatic, is to raise the standard of nutrition.

Niemeyer puts it thus : " Prophylaxis against consumption
requires, in the first place, that when an individual shows
signs of defective nutrition and a feeble constitution, especially if already he have given positive evidence of unusual
delicacy, with a tendency to diseases which result in caseous
products, he should be placed, if possible, under influences
calculated to invigorate the constitution and to extinguish
such morbid tendency."

Insufficient Diet.—" Among the influences by which a liabi-
lity to consumption is acquired, or by which a congenital
predisposition to it is aggravated, that of an insufficient or
improper diet stands first."

Impaired Nutrition.—Where the difficulty of insufficient

food does not meet us, we have often to face a condition of
chronic indigestion, aggravated, it may be, by improper dieting.   The quantity of the food may be ample, but the quality
may be such as cannot be assimilated by the weakened digestive organs.   The food may be too rich, too bulky, or too
coarse.   The result is the same in each case, namely, impaired nutrition.   In a word, the diet must be adapted to
the powers of the constitution at the time, and it must be
of the highest nutritive value that can be digested.

Small Meals.—In regulating the quantity and quality of
the food, especially during exacerbations of the malady, the
rule that a little thoroughly assimilated is better than a
larger amount only half digested, must be kept prominently
in view.   In the majority of cases, it will be found that the
quantity of food given at a time must be less than in health,
and consequently the intervals between the supplies must
likewise be shortened.   Consumptives find that a large meal
lies heavy on the stomach, and often gives rise to distressing
flatulence with acid eructations.

Distaste for Fat with Progressive Emaciation an Unfavourable Prognosis.—An unconquerable dislike to fatty articles
of diet is an almost constant accompaniment of this disease ;
but fat bacon, butter, and cod-liver oil can often be taken
without difficulty after the fat of fresh meat has become
distasteful.   The repugnance to fat seems to be associated
with an inability on the part of the system to digest fatty
foods, and coincident with this we usually find progressive
emaciation taking place.   Where the digestive functions are
fairly vigorous, and fatty articles of diet are not rejected,
the outlook is not nearly so bad as where the opposite
conditions exist.   In those cases, however, where dyspepsia
is severe and intractable, and especially where there is a
strong dislike to fats, with an almost entire inability on
the part of the system to digest them, a hopeful prognosis
cannot be given.   In all stages of the disease it is necessary
to keep as watchful an eye on the digestive functions as on

the state of the chest, and quiescence of lung symptoms, without improvement in nutrition, should not satisfy us.

Improved Nutrition the Object of all Treatment.—When we come to consider the details of dietaries for these patients, we shall see that they must vary from time to time according to the condition of the patients, but under all circumstances improved nutrition is our aim and object. When the patient is confined to his room or to bed, the greatest care should be bestowed upon the sick-room cookery.

Daintiness of Sick-Room Cookery.—Food must be served in the daintiest and most appetising ways, and not too much at a time. Usually it is well that the patient should not know what is coming ; a little pleasant surprise sometimes aids a doubtful appetite. Even when all these conditions are most favourable, appetite must not always be relied upon as a safe guide, for we not infrequently find that the system is able to digest and assimilate more food than the slender appetite would incline the patient to take.

Temporary Restriction of Diet.—While insisting thus strongly on the necessity for as liberal a diet as possible, it is not to be forgotten that there are times when the diet must be temporarily restricted; as, for example, when the tongue becomes furred, and when such symptoms as nausea, feverishness, and hurried respiration show themselves. Then it is time to limit the amounts of red meat and of stimulant, and altogether to simplify the dietary.

Milk Diet.—If gastric disturbance is severe, it will be controlled much more speedily by restricting the patient for the time being to a diet composed entirely of milk and prepared farinaceous foods, given in small quantities at frequent intervals. If we are very hard pressed, it may be necessary to give little or no food by the mouth, and to resort to the use of nutrient enemata, but the danger of diarrhœa being set up, or of its being increased if it has already appeared, renders feeding by enemata a difficult proceeding.

In some cases nutrient suppositories are well retained and are less irritating. Rubbing the body with oil may in extreme circumstances help to sustain life, but the discomfort is great, and in the case of cod-liver oil, at least, the smell is a great objection. Neat's-foot-oil and olive-oil are in respect of smell less objectionable, but they are not so effective.

Pure Milk.—Milk, good unskimmed country milk, should enter largely into the dietary of the consumptive. Some patients can drink milk warm from the cow and digest it well. One advantage of taking milk in this way obviously is that it cannot be skimmed, and patients, therefore, get the full nutritive value of the milk. Many, however, can take milk only when diluted, for example, with hot water or with an aërated water, or when boiled and with some farinaceous material in it.

Diluted Milk and Whey.—When difficulty is experienced in digesting milk in the ordinary form, it should be diluted with an aërated water or with hot barley-water, or it may be peptonised. Other plans for taking milk have been recommended, such as, for example, the "whey-cure;" but here the caseine being precipitated, and therefore lost to the patient, is a decided disadvantage; and whey should be given only when milk in other forms absolutely disagrees.

Koumiss.—Koumiss, the fermented milk of mares, was introduced into this country some years ago from Russia, and has obtained reputation as a beverage of considerable nutritive value in phthisis. Koumiss can be made at home from cow's milk, skimmed or unskimmed, the former closely resembling that made from mare's milk; the latter somewhat richer than the Russian koumiss, and therefore more nutritious where it can be digested.

Cod-Liver Oil.—Cod-liver oil has come to be regarded as an invaluable adjunct to the treatment of phthisis, and undoubtedly in many cases it is of the utmost use. It does not, however, suit all patients. Some have a great repugnance to it, and if this continues, with loss of appetite, the oil should not be persevered with. Although not the best form of fat, the readiness with which cod-liver oil is usually digested is a very great point in its favour, and it need not, as a rule, interfere with the taking of a plentiful supply of milk.

Times for taking Cod-Liver Oil.—The time at which cod-liver oil is taken is of importance. Some patients can take it easily at any time without its upsetting their digestion, but most find that to take it an hour or an hour and a half after meals is most agreeable. Theoretically, also, this is the best time, for digestion has so far advanced that the oil will not be long retained in the stomach, but will be passed on to be digested with the other fatty constituents of the food. Sometimes, for convenience' sake, the oil is taken immediately after meals, but where any difficulty in digesting it when taken soon after food occurs, the other plan should be tried. The quantity to be begun with should be small, as a rule not more than a teaspoonful twice a day, and this may gradually be increased to a dessert- or a table-spoonful twice daily. A few patients find that they can best take the oil in a single dose at bedtime.

Vehicles for Cod-Liver Oil.—As already said, many patients can take cod-liver oil easily; some seem to take it almost eagerly without admixture of any kind; but others prefer it shaken up in hot milk, or in a little fresh lemon-juice. Orange-wine or brandy and vegetable bitters, such as gentian and calumba, are also pleasant vehicles to some; and for children the syrup of the iodide of iron, or Parish's chemical food, answers well. From 10 to 20 minims of ether added to the dose of oil seem in some cases to increase its digestibility, and the same effect is produced by adding a small quantity of pancreatine powder or pancreatine wine, or by making it into an emulsion with liquor potassa, and disguising the taste by a drop or two of oil of cloves or of oil of cassia. Various combinations of cod-liver oil emulsified with malt extracts are sold, and form valuable preparations. It should also be remembered that in some exceptional cases the less refined forms of oil are better borne than the paler, purer oils.

Substitutes for Cod-Liver Oil.—Where cod-liver oil cannot be taken, mutton-suet, finely minced and dissolved in warm milk, is a useful substitute, and other oils, though far inferior to cod-liver oil in nutritive value, should be given where cod-liver oil causes too great repugnance, or where it manifestly disagrees.

Malt Extracts.—Next to cod-liver oil as additions to the ordinary dietary come malt extracts, which by their aid in the digestion of starches are oftentimes useful; and they may be given either plain or in milk, in effervescing waters, or in milk-puddings. In this connection, also, should be mentioned cream of malt, pancreatic emulsion, and similar preparations. As in other diseases where the digestive powers are feeble, prepared farinaceous foods are of great value, and the addition of liquor pancreaticus or of liquor pepticus helps greatly the complete assimilation of the food. In this way peptonised milk, peptonised gruel, peptonised beef-tea or soup (Recipes 74–78), and peptonised enemata (Recipe 79) are valuable resources.

Alcohol.—Bearing in mind the all-important fact that imperfect nutrition is the great difficulty against which we have to contend, we shall be guided in our use of alcoholic beverages by the effect they have upon digestion in each particular case. In early stages of the disease, and where food is well taken, they are, as a rule, not required to any great extent, or, if given at all, they should be combined with food, and the result carefully watched. If they do not unduly flush and excite the patient, and if the digestion of food is manifestly increased by their use, if more exercise can be taken with comfort, and a sense of general well-being is produced, we may be satisfied that they are doing good. If they produce opposite effects, the quantity taken is too large, or else they are altogether contra-indicated. The form of alcoholic stimulant is another important point, and one that cannot be decided on general grounds. Some patients do best on a glass of bitter ale with luncheon and with dinner, or with a glass of stout at these times. For others, some light wine, such as claret, hock, or chablis, is more suitable; others still can take only a small quantity of spirits, a tablespoonful of brandy in a beaten-up egg, or a like quantity of rum in milk, or of whisky in water; and in our selection we must be guided by individual experience.

Diet of Delicate Children.—In regard to children, the greatest attention should be paid to the diet of those who have inherited, or who by untoward circumstances have acquired, a feeble constitution, with more or less tendency to strumous manifestations, lest these be the forerunners of the phthisis that will show itself later in life. A consumptive mother should never nurse her own infant, and for such an one a good wet-nurse is far better than rearing by bottle. If, however, the bringing up of the child by bottle be unavoidable, it should, if possible, be reared in the country, and, failing that, great pains should be taken to secure the milk of a healthy, country-fed cow.

Value of Milk.—During the early months of its life, milk must be its sole food, and, later on, should still be the staple of its diet. Its place should never be taken by " pap," as it too often largely is, but some of the prepared farinaceous foods, mixed with the milk, may be used so long as the quantity of milk itself is not diminished. As soon as the child is old enough to masticate, it may be allowed biscuits and bread and butter, and taught to chew them well, in order to aid the development of the salivary secretions and to encourage the drinking of milk. A little tenderly-dressed meat, some well-boiled, tender, green vegetables and stewed fruits, all of which carry with them useful salts, should early enter into the dietary of such children.

Hygienic Measures.—It need hardly be added that every means to improve the appetite and invigorate the constitution generally should be fully carried out. Plenty of fresh air in rooms, an open-air life as far as that can be secured, freedom from continuous study, and removal, when it is possible, to such a climate as will permit of many hours of the day in winter being spent out of doors, are most valuable accessories. All this applies equally, and even more strongly, if possible, to the period of growth succeeding childhood in girls and youths of delicate constitution. An open-air life, with plenty of every kind of healthful exercise, an abundant but simple diet, and limited hours of study, give such young people the best chance of escape, and often work wonders in building up the constitution.

Dietary.—A diet such as the following will be suitable :—

The first thing in the morning, a cupful of warm milk, or of milk warmed.

Breakfast, 8 A.M.—Porridge (oatmeal, whole meal, or hominy) and good milk ; a slice of fat bacon or an egg, with brown bread and butter, and a cup of cocoa.

11 A.M.—A cup of milk and a small biscuit.

1.30.—Any plainly-cooked meat, mashed potato, and well-boiled green vegetables, simple pudding, and stewed fruit.

4.30.—A cup of cocoa and a slice of brown bread and butter.

7 *o'clock.*—A cupful of bread and milk, or of milk with bread and butter, and an egg if there has been plenty of air and exercise.

In some cases it will be desirable to give from one to two teaspoonfuls of cod-liver oil an hour or so after breakfast, and again after the evening meal ; or to give a teaspoonful of malt extract with the midday and evening meals.

The diet we should advise for an adult threatened with consumption is as follows :—

Early in the morning a cup of milk with the chill taken

off, or a cup of milk flavoured with tea, and a slice of thin bread and butter.

Breakfast, 9 A.M.—Fat bacon, fish, or eggs, or eggs and bacon, with brown bread and butter, *café au lait,* or cocoa and milk.

1 *o'clock.*—Chicken, game, or light meat, mashed potato, and green vegetables, such as spinach or tender cabbage. Milk-pudding, or shape or custard, with cooked fruit. A glass of light bitter ale, or a couple of glasses of light wine.

4.30.—A cup of tea, half milk, bread and butter, or plain cake.

7 *o'clock.*—A meal like the midday dinner.

Before going to 'bed, a glass of milk, or of prepared farinaceous food with milk, with or without a tablespoonful of spirit, as the circumstances may indicate.

Modification of Diet.—This diet must be modified to suit each special case. For example, if the patient cannot take the early morning milk without losing his appetite for breakfast, let him have an early light breakfast, and at 11 A.M. a cup of soup or of broth with toast or a biscuit. Again, if the evening meal is preferred early, let it be taken say at 6.30, and then give about 9.30 a cup of rusks and milk or of arrowroot, with a tablespoonful of spirit. Arrowroot and brandy will answer well when there is any tendency to diarrhœa; and when such is the case, the green vegetables and fruit will have to be omitted for the time being. Moreover, if there be much dyspepsia, the diet may have to be still further simplified, and divided into smaller meals at shorter intervals.

Where the disease has become established, but there is an absence of fever, the following will serve as a type on which to work out a dietary suitable for the special case :—

7 A.M.—A cup of milk, with a tablespoonful of rum or other spirit in it, or a cup of milk with a little tea in it, and a slice of thin bread and butter.

Breakfast, 8.30.—Eggs, or egg and bacon, or fish with a little melted butter, or porridge and cream, bread and butter, a cup of *café au lait* or of cocoa.

11 A.M.—A small cup of soup or of beef-tea, thickened with some farinaceous material.

Early Dinner, 1 *o'clock.*—Any tender well-cooked meat, especially mutton, chicken, game, with mashed potato, and green vegetables of the tenderer sorts; any simple pudding or stewed fruit.

Beverage.—Half a pint of light bitter ale or of stout, or a couple of glasses of claret or of chablis.

4.30.—A cup of cocoatina, or of tea with plenty of milk, and a slice of bread and butter, or cake or plain biscuit.

7 *o'clock.*—A light but good meal of fish or chicken, or sweatbread or tripe, with dry boiled rice or bread; plain cream or jelly, a couple of glasses of light wine, or two tablespoonfuls of spirit in a little water.

9.30–10.—A cupful of prepared farinaceous food, or of arrowroot with a tablespoonful of brandy in it.

Hectic Stage.—During the continuance of fever, and in the later stages of consumption generally, the diet, as has already been pointed out, must be greatly simplified and limited mainly to liquid food. Moreover, since the quantity taken at a time will be small, the intervals between the supplies must be short—two hours, or even less. It is often well to adhere to the hours of ordinary meals, and to give the other supplies as intermediates. Although there may not be much difference in quantity or even in the kind of food, still it is less monotonous for the patient. Milk with an egg beaten up, or some malt extract or spirits in it, mutton or chicken broth, or beef-tea, thickened with some farinaceous food, lightly-boiled eggs, sweat-bread, chicken, game, and fish, with custard, jellies, and plain puddings, will give variety to the meals. It is often well to peptonise the milk, or to dilute it with barley-water, or even with an aërated water. Alcohol is most useful at this stage of the disease, and generally agrees best in the form of old spirit or of dry sherry, marsala or of a light port.

Thus, early in the morning (an hour and a half to two hours before breakfast), a glass of milk warm, with a table-spoonful of milk or of brandy in it.

Breakfast, 8.30.—A cup of weak tea with milk, or of *café au lait* with an egg beaten up in it, with some thin bread and butter, or toast and butter, or rusk, or, if the patient prefers it, an egg lightly boiled or poached on toast, with a small cup of tea filled up with cream.

10.30 A.M.—A cup of "whole" beef-tea, or of good soup or broth, with a finger of toast.

Luncheon, 1 *o'clock.*—A bit of chicken or chicken *panada*, or sweetbread, or white fish with rice or bread, a spoonful of blanc-mange, or jelly, or plain cream. A glass of wine or a tablespoonful of spirit in a claret-glassful of water.

4 *o'clock.*—A cup of tea as at breakfast, or of cocoa and milk (preferably peptonised), or of milk and Mellin's food.

Dinner, 6.30.—A meal like luncheon, but of course varied as much as possible. Chicken, rabbit, sweetbread or bird, or fish, such as sole or whiting, with a little melted butter; a spoonful of custard or of milk-pudding, a glass of wine or a tablespoonful of spirit as at luncheon.

9.30–10.—A cupful of arrowroot, or of some prepared farinaceous food, with milk and a tablespoonful of brandy. Some peptonised milk or prepared food, kept warm under a cosey or heated by a spirit lamp, should be within reach during the night.

Diarrhœa.—Diarrhœa, occurring in the course of phthisis, calls for careful attention and demands strict dieting. It will often be sufficient to exclude from the dietary, for the time being, fruits, vegetables, brown bread, broths, and soups; to limit the amount of meat, and to give chiefly milk, pre-pared farinaceous foods, arrowroot, and brandy, taking care that the quantity given at a time is small, and that nothing is taken either hot or very cold. For further details of treatment, the reader is referred to the section on Diarrhœa (p. 34).

Constipation.—Constipation is sometimes troublesome in phthisis, and it should be dealt with as directed in a special section dealing with that subject (p. 33). It must, however, be remembered that there almost invariably exists a great tendency to irritability and catarrh of the bowels, not to speak of ulceration, and that therefore the greatest care should be exercised, lest by treating constipation too vigorously, the opposite condition of diarrhœa be set up.

ASTHMA.

In most cases of asthma, apart from those that merit the distinctive title of "peptic," dietetic measures greatly aid medicinal treatment.

Digestive Disorders.—Even if some error of diet be not the immediate exciting cause of the attack, a disordered condition of the digestive organs sooner or later ensues, and greatly aggravates the patient's state. Asthmatics are more or less dyspeptics, and they must be treated as such. Their diet should be carefully regulated. Bulky foods should be avoided, and not only the kind of food taken, but the times of eating must be specially arranged. Breakfast and mid-day dinner should be the chief meals, and no such thing as a regular late dinner or supper is to be permitted.

Small Evening Meals.—A small early evening meal of the simplest sort only can with safety be allowed. As a general rule, bulky green vegetables, much fruit, and any large quantity of starchy foods, which might distend the stomach and bowels, are to be carefully excluded from the dietary of the asthmatic, and he should be warned against taking much liquid of any kind with meals. Tea, effervescing beverages, and soups are to be taken, if at all, in the strictest moderation, while of meats, mutton, chicken, game, and fish are to be preferred to the less digestible forms.

Alcohol.—Alcoholic stimulants are necessary in cases where there is much depression and weakness, and where they seem to aid digestion, but their use is to be most carefully watched and limited. Strong coffee is so useful to many asthmatics at the time of a paroxysm, that its habitual use is not recommended, since this would tend to lessen the benefit derived from it at those times.

What has just been said regarding the diet of asthmatics generally applies with even greater force when they happen to be placed in circumstances that notably favour an attack.

Dietary.—The following may be taken as the type of a diet embodying the important points suggested above :—

Breakfast.—Fresh fish, or chicken, or game, or bacon, or egg and bacon, with toast or brown bread and butter, and one cup of cocoa nibs.

Early Dinner (four to five hours after breakfast).—A plain substantial meal of two courses. Fish and meat, or meat and pudding; any tender well-cooked meat with a little green vegetable well boiled and mashed, plain pudding, or shape and stewed fruit.

Beverage.—A glass of water, with or without a glass of some light wine, or a small glass of bitter ale.

Evening Meal, 6 or 6.30.—An egg lightly boiled and a slice of bread and butter, or an egg beaten up in milk and a biscuit, or a cupful of prepared farinaceous food, such as Mellin's or Benger's.

Modification of Diet.—It may be difficult at first for the patient to get accustomed to having no more substantial food in the latter part of the day, but the benefit usually derived from a plan such as the above will reconcile him to the change, and a sufficiency of nutriment being taken at the morning and midday meals, the light evening repast will prevent a feeling of exhaustion and sleeplessness. At first, however, it may be necessary with some patients to let them down easily by permitting a cup of cocoa or weak tea to be taken about five o'clock, and a small evening meal like the

above given at seven, assuming that the patient does not go
to bed till ten or later.   The evening meal, however, should
in any case be taken at least three hours before bedtime.

Pulmonary Emphysema.

While the treatment of pulmonary emphysema resolves
itself largely into the question of treating the complications
and secondary conditions that sooner or later manifest themselves, the diathesis of the patient should be kept prominently
in view, and every means should be employed to improve
the general health.   In this way degenerative changes may
be retarded or checked, even if past mischief cannot be
repaired.

Gout.—If the patient be gouty, he must be dieted accordingly, only it should be borne in mind that in permissible
things he is not to be too greatly stinted.

Cod-Liver Oil.—In addition to ordinary simple nourishing foods, cod-liver oil may often be given with advantage,
especially in cold weather.

State of the Lungs.—Attention to the state of the system
generally is most important, and while the condition of the
digestive organs, the action of the skin and bowels, the influence of fresh air with gentle exercise, and the effects of
climate, must each and all be kept in view, the state of the
lungs themselves must not be forgotten or overlooked.   With
usually a considerable decrease in size, there is always diminished activity of, and retarded circulation throughout, the
lungs, and consequently a corresponding diminution in the
processes of oxygenation.   Add to this the fact that the
patients are restricted in muscular activity, and it is evident
that the diet must be adapted to the altered circumstances
if the patient is to be saved unnecessary discomfort and
increased dyspnœa.

Diet as for Asthmatics.—In many respects the rules laid
down for the diet of asthmatics will be found suitable in

the cases now under consideration. In most points the diets will be similar, and more particularly in those instances, and they are not a few, where there exists a marked tendency to spasmodic dyspnœa. The diet should in all cases be nutritious, and it should be somewhat concentrated, although at the same time due regard must be had to variety and to promoting the action of the bowels. Care must be exercised that the stomach and bowels be not distended by bulky articles of diet, inasmuch as such distention is a fruitful source of dyspnœa. The meals must be small, and the main portion of the food should be taken early in the day, no late dinners or heavy suppers being permitted. (See Asthma, p. 108.)

CHRONIC BRONCHITIS.

As regards the diet in cases of chronic bronchitis, it may be said in a general way that lowering measures are not called for, and that the patient requires a liberal supply of light, nourishing food.

Diathesis.—The question of the patient's diathesis must likewise be taken into account. A gouty tendency is frequently present, and must not be overlooked in arranging the dietary. Moreover, the condition of the liver and of the kidneys demands consideration ; and lastly, a very important point is the question of cardiac complications. In cases where the heart is weak, or where there is chronic valvular insufficiency, the directions given for the dietary of such patients (see chapter on Chronic Heart Disease) should be followed.

Dietary.—Regarding uncomplicated cases of chronic bronchitis, while the food should be simple, light, and easy of digestion, it should not be bulky, and the amount of liquid in the dietary should be small.

Dry Diet.—A somewhat dry diet helps in many cases to diminish abnormal secretion, and for the details of such a diet the reader is again referred to the section on chronic cardiac diseases.

This applies also to cases of bronchiectasis.

PLEURISY WITH EFFUSION.

Active measures for promoting the absorption of the fluid are usually resorted to, but it is frequently forgotten that great benefit may accrue from a reasonable limitation of the amount of fluid that the patient is allowed to consume. In dieting a case of this nature, the facts to be kept chiefly in view are the necessity that exists for maintaining the patient's strength, and for promoting the absorption of the fluid. Regarding the latter point, it is scarcely necessary to point out that free elimination by the kidneys, skin, and bowels is of comparatively little use if the amount of fluid taken into the system be not diminished.

Dry Diet.—This necessity for limiting the liquid part of the diet is not usually made prominent by writers on the treatment of pleurisy, and it has not received the attention it deserves. Patients find it difficult at first to limit the quantity of liquid taken in the twenty-four hours ; but if the plan be not pushed too much at first, and if the object of the restriction be pointed out, the necessary curtailment can usually be secured without difficulty, and, after the first day or two, without discomfort.

Diaphoretics, Diuretics, and Purgatives.—It has been recommended that this treatment by dry diet should be tried only after the use of diaphoretics, diuretics, and purgatives has failed, but there is no reason why in most cases it should not be tried first, or in combination with other means, and not as a last resource before tapping the chest. Diuretics are uncertain in their action, and strong purgatives of the drastic class are trying to the patient, besides being in some cases absolutely contra-indicated ; but free diaphoresis, as, for example, by jaborandi, is undoubtedly useful in many instances.

It must not be understood that in advising a dry diet the total exclusion of liquids from the dietary is advocated. Such a diet is extremely irksome and trying to the patients. By some it cannot be borne ; in fact, they would rather have thoracentesis at once performed, and, under proper precaution, that operation may be considered perfectly safe. Even after withdrawal of the fluid, however, a moderately dry diet is useful, especially in cases where there is a tendency to re-accumulation.

Dietary.—In detail the diet recommended is as follows :—

8 A.M.—An egg or a piece of bacon with toast, and one small teacupful of cocoa, or a cupful of bread and milk (the milk being about a teacupful).

10.30–11.—Three or four little sandwiches of pounded chicken and bread, with a small glass of wine.

1–1.30.—Fresh white fish, or *panada* of chicken or game, or a piece of boiled chicken with a spoonful of mashed potato and a little well-boiled green vegetable. Following this, if the appetite is good, a few spoonfuls of any light pudding.

Beverage.—From one to two glasses of light wine, such as hock, chablis, or claret.

4.30–5.—A small cup of milk or of cocoa, or of weak tea, with a slice of thin bread and butter.

7–7.30.—A meal like the midday one.

10.30.—A teacupful of arrowroot, with a dessert-spoonful of brandy in it, or a teacupful of Mellin's or of other farinaceous food.

In long-continued cases, and in weakly persons, cod-liver oil is a valuable addition to the dietary.

CHAPTER XII.

DISEASES OF THE HEART.

GENERAL CONTENTS : Chronic Valvular Heart Disease—Weak Heart—Fatty Heart—Aortic Aneurism.

CHRONIC VALVULAR HEART DISEASE, WEAK HEART, AND FATTY HEART.

IN cases of valvular disease of the heart, so long as compensation is maintained, the patient does not suffer from the valvular insufficiency, and he may even be perfectly unaware that his heart is affected at all. So soon, however, as the compensation becomes disturbed, the balance of the circulation is destroyed, and the mechanical effects of such a condition manifest themselves.

Symptoms.—The pressure in the systemic arteries becomes less, and that in the veins greater, causing in a very short time a back tide on the lungs, the liver, the stomach, and the whole portal circulation.

Flatulence, with Inactive Liver and Bowels.—The patient complains of a feeling of fulness and weight in the stomach. The liver is inactive, and the bowels are very sluggish in their action. Flatulence is often very troublesome ; the urine, somewhat diminished in amount in the earlier stages, is loaded with urates; in the later ones it is scanty and contains also albumen. How can we in such a case help our patient by the regulation of his diet ?

Selection of Food.—Two things are evident : the food must be easy of digestion, and it must not be bulky. If the food be slow in digesting or bulky in character, flatulence, which we have already said is often a very troublesome symptom in such cases, will be increased, and as a direct result there will be impeded action of the heart, with all its distressing consequences.

Directions.—Food, then, must be given in the most easily-digested forms, in comparatively small quantities, and at moderate intervals.

(*a.*) *Early Stage.*—If the patient has sought advice early, before the pathological conditions above mentioned have become marked, it will suffice for his comfort and well-being that we direct him to divide his supplies of food into four small meals a day, and to eat red meat only once a day, and that at the principal meal; to make the evening meal a light one; to limit the quantity of liquid taken at meals, as well as the whole amount taken in the twenty-four hours; and to avoid everything difficult of digestion. Under this head we shall class cured meats, pork, salted or smoked fish, shell-fish, pastry, rich puddings, much or strong tea and coffee, and all effervescing beverages.

Later Stages.—When the early stage has been passed before the patient comes under our care, we shall have to select carefully the most suitable articles of diet for his particular case, and we must with equal care regulate the times of eating, and the quantity of food taken at a time. The meals must be small and the intervals short. Some food must be given before the patient undergoes the fatigue of dressing, and therefore we order a light breakfast in bed.

8 A.M.—A lightly boiled egg or a small piece of fresh white fish, with two small slices of dry toast and a little butter; one small cup of tea, not strong, and with plenty of milk in it. If the egg cannot easily be taken in any other form, it may be beaten up in the tea; and should there be much exhaustion, it is well to add a teaspoonful of brandy.

11 A.M.—A teacupful of strong soup with a finger of toast, or half a glass of port wine with a small biscuit, will

dispel the fatigue apt to follow upon the exertions of the morning toilette.

1–1.30 (*this should be the chief meal of the day*).— The lighter meats must be chosen, roast or boiled mutton, about four ounces, or a mutton chop or a plain cutlet, may be varied by the occasional introduction into the dietary of a like quantity of fresh minced meat, or the wing of a chicken, or calf's head or pig's cheek. Green vegetables are usually contra-indicated on account of the flatulence, which they increase, but a little mashed spinach, French beans, cauliflower, or vegetable marrow will often be well borne. This same tendency to flatulence leads such patients to indulge rather freely in mustard, pepper, and other condiments; but great care must be exercised in their use, owing to the risk of gastric catarrh. Briefly, then, dinner will consist of a slice of roast or of boiled mutton, or a mutton chop, or a slice of pig's cheek or of calf's head, with a spoonful of well-cooked vegetable. Following this may come a few spoonfuls of any plain milk-pudding, seasoned with cloves or cinnamon, or a baked apple with a spoonful of custard.

Beverage.—From a dessert- to a table-spoonful of old spirit in a claret-glassful of water, sipped towards the close of the meal.

5 *o'clock.*—The afternoon cup of tea will be looked forward to, and need not ordinarily be forbidden, but it must be one cup only, with a slice of toast and butter, or of stale bread and butter.

7.30 P.M.—Dinner may consist of an egg lightly boiled or poached on toast, or a little bit of fresh white fish, such as sole, whiting, or plaice; or tripe in white sauce with bread; also sweetbread, simply cooked (not braised), makes a pleasant variety (Recipes 48–51).

Beverage.—A tablespoonful of old spirit in a little water, as at the midday meal.

A full evening meal always leads to trouble in these cases, causing discomfort, and even distress, with difficulty of breathing in the night; hence the necessity of the small and early supper. It will, however, often conduce to the patient's comfort, and aid his sleeping, if he takes a teacupful of cornflour, or of arrowroot, or of one of the prepared farinaceous foods, with a tablespoonful of brandy stirred into it, just before settling for the night, or a little milk warmed, and with the same amount of brandy in it, may be sipped at that time.

Advanced Stage.—When cases of chronic heart disease come before us in an advanced stage, the question of their dietary is a serious one, but still great help and much relief may often be given even to such patients, by dieting only.

Dropsy.—The difficulty of breathing may be distressing; the loading of lung, liver, and other organs troublesome; dropsy may have supervened; and yet under a suitable dietary these symptoms may be greatly relieved. The patients will tell us how much difficulty they have in digesting any food, how great their discomfort from flatulence is, and that thirst continually troubles them.

Gastric Catarrh.—It may be that gastric catarrh is manifestly present, and this will add to the difficulty of our task. On questioning the patients as to the quantity of liquid taken within the twenty-four hours, we find, as a rule, that they are consuming much more than in health, and that at the same time a small quantity of urine is passed; the skin is inactive, and the bowels sluggish. Little wonder, then, that the lungs and the abdominal organs are all congested, and that the anasarca is increasing.

Purgatives and Diuretics.—Purgatives, producing watery evacuations, and diuretics, largely increasing the flow of urine, are most valuable in getting rid of a proportion of the fluid; but if the quantity of liquid ingested be not lessened, the strain on the excretory organ is enormous.

Value of a Dry Diet.—Hence the value of a dry diet in these cases. If the circumstances are explained, so far as possible, to the patient, and his co-operation secured, it is

wonderful how quickly he gets accustomed to the limitation
put upon his supply of liquid. Thirst becomes gradually
less trying, and the feeling of relief arising from the subsi-
dence of the urgent symptoms gives him courage to persevere.
Light, nourishing food in small quantities, and not too liquid
forms, at frequent intervals, must here be the rule. There
is great need for keeping up the patient's strength, but he
cannot digest much at one time. A full meal, if retained
by the catarrhal stomach, would only lie like a weight
upon him, increasing flatulence, and embarrassing the heart's
action.

Dietary.—It will be difficult to dispense with the morning
cup of tea, so refreshing to the patient after a night of somewhat broken rest; therefore at

8 A.M.—Let the first meal consist of a small cup of tea
with milk in it, and a slice of dry toast and butter. Here,
again, the egg beaten up, with from one to two teaspoonfuls of brandy in the tea, will be the most supporting and
easiest-taken form of nourishment. Variations on this meal
will be a cupful of bread and milk, or a cup of cocoa with
toast soaked in it, and the directions for taking it slowly by
teaspoonfuls hold good here equally as in the case of ordinary dyspeptics.

10.30 A.M.—A teacupful of good soup or strong beef-tea,
thickened with some farinaceous material, sago or vermicelli,
or with fingers of toast dipped in it. The nutritive value
and ease of assimilation will be increased by the addition of
peptones of beef. In cases of great prostration, from one to
two teaspoonfuls of brandy may have to be given with this
supply of food.

1.30 P.M.—A plain *purée* of meat or *panada* of chicken,
or pounded meat, with bread crumbs and rice (the cooked
meat not to exceed three ounces in weight). From one to
two tablespoonfuls of brandy will be allowed with this meal.

4.30 P.M.—A teacupful of shortly-infused tea or of cocoa,
with a slice of thin bread and butter, or toast, or rusks.

6.30 P.M.—A lightly-cooked egg, a teacupful of beef jelly or of turtle-soup, flavoured with a dessert-spoonful of old madeira, or a slice of tripe, or an ox palate, with bread. Two tablespoonfuls of brandy must be repeated at this meal.

9 *o'clock.* —An egg beaten up in a little milk and brandy, to the amount of one tablespoonful, or two teaspoonfuls of Brand's essence of meat, with the like quantity of brandy. This may be repeated during the night, or a little warm milk, with a dessert-spoonful of brandy, may be given.

AORTIC ANEURISM.

A sufficient number of cases of aortic aneurism in which improvement has followed treatment by rest and diet have been reported, to render it incumbent on every one placed in charge of such a patient to give this treatment a fair and exhaustive trial.

We must not run away with the notion that it is simply a question of reducing the diet to the smallest amount of food upon which life can be supported. The patient must not be starved, or else, while the aneurism is being cured, he will be killed. What has been termed the "*vita minima*" plan has not succeeded, and has probably done more harm than good. If we consider for a moment the importance of not impoverishing the blood too greatly, we shall see that this must be so. Impoverished blood will lead only to anæmia and dropsy, and will thus defeat the very object we have in view, viz., the deposition of fibrin from the blood within the sac of the aneurism.

Points of the Diet.—The chief points to be borne in mind are these: (1) the diet must be as dry a one as possible; (2) it must be largely nitrogenous; and (3) it must be so distributed over the day as to minimise the risk of even temporary plethora. To meet the first of these conditions, we reduce the quantity of liquid given in the twenty-four hours as far as can be borne by the patient. To make the diet supporting, and with a view of aiding the coagulation of the blood within the sac of the aneurism, we make the proportion of nitrogenous food greater than we should consider in other circumstances requisite to balance the dietary. Any approach to acceleration of the circulation being most undesirable, food will be given in small quantities at a time, and nothing like a full meal allowed. Except in cases where there is much prostration, alcohol should not enter into the dietary. In cases of great prostration we must, however, allow a very limited amount, and the safest form will be 2 oz. of claret or of a light burgundy with the food once or twice a day.

The patient must assume a horizontal posture, and be willing to remain at perfect rest in that position for a period of from two to three months. The directions then would be the following :—

Breakfast.—An egg or 2 oz. of fish ; 2 oz. of bread and butter; 3 oz. of cocoa or of milk.

Dinner.—Two oz. of meat ; 2 oz. of potato or bread, or dry boiled rice ; 3 oz. of water.

Tea.—One oz. of bread and butter ; 2 oz. of cocoa or milk.

Supper.—One egg ; 1 oz. of bread and butter; 2 oz. of milk. Total, say 12 oz. of solids, 10 oz. of liquids.

Mr. Tufnell's Plan.—The plan proposed by Mr. Tufnell, and which bears his name, directed that the patient should be kept in bed and at perfect rest in the horizontal posture for a period of from eight to thirteen weeks, according to the effect produced upon the aneurism, every care being taken to avoid movement in bed, special arrangements being made that the bowels and bladder might be relieved without altering the position of the body. The only medicines used were occasional opiates and laxatives. The following was the diet prescribed :—

Breakfast.—Two oz. of white bread and butter ; 2 oz. of cocoa or milk.

Dinner.—Three oz. of meat; 3 oz. of potato or bread; and 4 oz. of water or claret.

Supper.—Two oz. of bread and butter, and 2 oz. of milk or tea ; giving a total quantity of solids 10 oz., and of liquids 8 oz.

This plan in Mr. Tufnell's hands resulted successfully in ten cases, and although a rigid one, might be carried out with the co-operation of an intelligent patient and a reliable nurse. In many cases, however, it will be found necessary to modify it to the extent allowed in the former dietary.

CHAPTER XIII.

ACUTE AND CHRONIC RHEUMATISM.

ACUTE RHEUMATISM (*Rheumatic Fever*).

WITH some important modifications to be afterwards emphasised, the diet in acute rheumatism is the same as in other acute febrile conditions, and the dieting of the patient is a very impo:tant part of the treatment.

Early Stage.—In the early stages of the disease it is not, as a rule, difficult to restrict the patient to a suitable diet; the difficulty rather may be to get enough nourishment taken; but with the return of appetite during convalescence, it is hard for him to believe that a good supply of butcher's meat will not make his progress towards recovery more rapid.

Diet must be Non-Nitrogenous.—As a matter of fact, however, we know that a return to solid food, and more especially the giving of meat too soon, is almost certain to be followed by a relapse. In acute rheumatism, more than in most other diseases, the system is loaded with waste products, the results of imperfect assimilation, and the powers of the digestive organs are very seriously impaired.

Milk with an Alkali.—So long, therefore, as the symptoms are acute, small quantities of milk with an alkali, or with an alkaline water, such as potash, soda, vichy, or limewater, should form the main part of the dietary, and

122

besides these a little beef-tea, or chicken-tea, or mutton broth only may be allowed. The quantity to be given at a time is a very important matter, for it must not be forgotten that any considerable amount of even liquid nourishment may materially affect the heart's action, in the then usually embarrassed condition of that organ. Two to three ounces every hour, or every two hours, are better than double that quantity at longer intervals.

Relief of Thirst.—Thirst may be relieved by sips of an aërated water, and the addition of a little lemon-juice makes it more palatable, and does not seem to be contraindicated. If there be diarrhœa, lime-water with the milk is useful.

Farinaceous Substances.—As the temperature falls and the acute symptoms subside, vegetable soups (Recipe 41, 41A), bread, and other farinaceous substances may be gradually added to the dietary; also gruels, milk-puddings, malted foods, arrowroot, cornflour, rice, and the yolk of an egg beaten up with a little milk and a spoonful of brandy. As convalescence proceeds, next will come fresh white fish boiled, sweetbread, or chicken once a day, and the patient should be kept to these things until some days after all rheumatic symptoms have entirely disappeared. Relapses can not uncommonly be traced to an unwise haste in returning to ordinary diet. It need hardly here be pointed out, that any chill, besides risking a return of the inflammatory symptoms, is dangerous, inasmuch as it seriously interferes with the nutritive processes, and thus materially retards convalescence.

Alcohol.—As regards alcohol, it may be said that it is required in most cases of acute rheumatism, indeed in all but those slight cases where the symptoms are neither severe nor long continued. Feebleness or inequality of the pulse, even without irregularity, is an early indication for ordering alcohol; and if attended to, the more serious prostration and tremulousness of muscles that often follow may be averted. The quantity, however, should be strictly regulated, and the effect upon the heart's action carefully observed. It is best to begin with a small allowance, say two ounces of brandy in the twenty-four hours, given in divided doses of a teaspoonful, a dessert-spoonful, or a table-spoonful, according to the age of the patient. In cases of marked cardiac failure, alcohol must be given liberally in larger doses, repeated as often as may seem necessary. The total amount ranges as high in some severe cases as twelve, eighteen, or even twenty ounces in the twenty-four hours.

CHRONIC RHEUMATISM AND RHEUMATOID ARTHRITIS.

The diet must be liberal and most nutritious, but at the same time it must be light, and the digestive organs should not be burdened with any excess. Whilst, therefore, food must be adapted to the circumstances and digestive powers of each individual patient, the dietary should contain as much fat and fatty matters as possible. All rich and complicated dishes should be avoided, and also all cured meats, dried fish, pies, pastry, and sweets should be omitted from the patient's list.

He may eat at *breakfast* fat bacon or fat ham, brown bread and butter, and have a cup of cocoa and milk.

Luncheon.—Fresh fish with melted butter, or chicken, tripe, or calf's head, mashed potato, and well-boiled green vegetables.

Afternoon Tea.—A cup of tea or of cocoa with milk, and a slice of thin bread and butter.

Dinner.—If fish has been taken at luncheon, any plain meat, once cooked, may be taken at dinner with a little mashed or grated potato, and some well-boiled green vegetable. Any plain pudding or cream, or stewed fruit and cream.

Alcohol.—A small allowance of alcohol is usually beneficial, and the most suitable form for these patients is two tablespoonfuls of spirit in plain or in effervescing water.

Cod-Liver Oil.—Cod-liver oil, malt extracts, or a combination of the two, cream, or so-called cream-of-malt, are useful adjuncts to the dietary.

When sufficient nourishment is not taken at the regular meal-times, the dietary may be usefully supplemented by giving a cupful of prepared farinaceous food at bedtime or during the night. Old people frequently require the latter.

CHAPTER XIV.

ALCOHOLISM.

CHRONIC ALCOHOLISM.

THE furred tongue and ethery breath, the watery eye and shaky hand of the man who habitually indulges too freely in alcoholic stimulants, tell their own tale, and hardly need the history of disturbed sleep and other nervous distresses, or of morning sickness and different gastric disturbances, to confirm the diagnosis.

In all cases of chronic alcoholism, the digestive powers are more or less weakened, the nervous and muscular systems enfeebled, and the nutritive functions impaired.

Dietary.—(*a.*) *In Severe Cases.*—When severe cases of long standing come under our observation, marked alterations in the structure of the viscera have already taken place. Fatty degeneration and fibroid contraction are the characteristic changes met with in the different organs. In the stomach chronic gastric catarrh, with fibroid degeneration and atrophy of secretive glands, is prominent. Fatty degeneration of the heart and blood vessels results in feebleness of the circulation and a marked tendency to local congestions ; similar changes in the liver and kidneys, advancing to cirrhosis, seriously interfere with the functions of these important organs. Nor is the nervous system exempt from changes of a like kind. The tissues of the whole body are pale, flabby, and fatty or fibroid ; the functions of the different organs are not actively performed, and in these circumstances absorption and assimilation, on the one hand, are very imperfect, while, on the other hand, the elimination of waste products is much hampered. Nutrition, therefore, is almost at a standstill. These patients require nourishment in considerable amount, but they have no appetite for food, and, as we have already seen, their digestive powers are but feeble. They consume only a small amount of food, and they are not able fully to digest what they do take.

Conditions of Recovery.—The conditions necessary to their restoration to health are, *first*, to leave off alcoholic stimulants altogether ; and, *second*, to take as much simple nourishment in regulated quantities as we find they can digest. The diet, therefore, must be adapted, as far as it is possible to do so, to the degree of structural change that has taken place. The more advanced the condition, the more simple the diet. Food of the simplest kinds must be given in small quantities at short intervals. Remembering that the early morning is the worst time with such patients, that then they feel depressed and good for nothing, an early light breakfast should be given them in bed.

Breakfast (say 8 *o'clock).*—A cup of weak tea with an egg beaten up in it, and a slice of toast, or a cupful of revalenta arabica (Du Barry's).

Second Meal, 10.30.—A cupful of good meat soup, thickened, or a cup of peptonised milk, warmed (Recipe 74), with toast or rusk.

Early Dinner, 1 *o'clock.*—A tablespoonful of pounded meat on toast, or a small teacupful of fresh meat juice, with fingers of toast. A couple of spoonfuls of sago, custard, or tapioca pudding. *Beverage*—Half a tumblerful of milk and soda-water (equal parts).

Afternoon Tea, 4.30.—A cup of weak tea or of cocoa nibs, with rusk or a plain biscuit.

Evening Meal, 7 *o'clock.*—A cup of soup or of mutton broth, as at the forenoon meal.

9.30 *or* 10.—A cup of peptonised milk and a slice of thin bread and butter.

Food During the Night.—In case of wakefulness there should be within reach, and arranged so as to be readily warmed, peptonised milk. The ordinary Etna or spirit lamp will be convenient for this purpose.

(*b.*) *Slighter Cases.*—In slighter cases we may have only the characteristic mawkish breath and the muscular tremors, or other commencing nervous symptoms, indicating the patient's danger and the necessity for reforming his mode of life; and in these milder degrees of the condition, if we succeed in inducing him to leave off every form of alcoholic stimulant, and to take plenty of plain, simple food, rapid restoration to health may be anticipated. We should diet such a patient in the following manner:—

Breakfast.—A lightly-boiled egg, or a little bit of white fish or of bacon, or a light savoury omelette; toast, or bread and butter; a cup of weak tea or of coffee and milk.

Early Dinner, 1 *o'clock.*—A slice of roast mutton, or the wing of a chicken, or a plain cutlet, with a spoonful of mashed potato and some well-boiled green vegetable. A few spoonfuls of any plain milk or bread pudding, with occasionally some stewed fruit. *Beverage*—A glass of plain or aërated water, to be taken slowly towards the close of the meal.

Afternoon.—A cup of tea with milk may be allowed, with toast or rusk.

Supper, 7.30.—A light meal of fresh fish, or of bird, or of calf's head, or of tripe, with dry-boiled rice or bread. *Beverage*—Small glass of milk and soda-water.

ACUTE ALCOHOLISM (*Delirium Tremens*).

Anorexia is always a prominent symptom in cases of acute alcoholism, especially at the beginning of the attack. If the subject of acute alcoholism has been a chronic drinker, and

the pathological changes briefly enumerated under the head
of Chronic Alcoholism have been established, his powers of
digestion are so enfeebled that alimentation becomes a matter
of extreme difficulty. The condition may, however, arise
occasionally in more healthy subjects who have not been
habitual drinkers, and whose constitutions, therefore, are not
so far gone as in the case of those who are saturated with
alcohol. In both classes the question of alimentation is
of vital importance, but in the latter it will probably be
managed far more easily and successfully than in the former.
In the latter, moreover, there is no fear of evil results following upon the omission of alcohol in every form from the
dietary. As soon as the patient can be induced to take a
good supply of nourishing food, the depths of his depression
have probably been reached, and there is every hope of his
soon obtaining refreshing sleep. In beginning the treatment,
small quantities of food must be given at intervals of two
hours, or oftener, if the quantity taken at a time be very
small and the patient extremely restless.

Dietary.—A cupful of good soup, mutton-broth, or chicken-
tea, thickened, should alternate every two hours with a
cupful of milk, plain or peptonised.

An egg beaten up in weak tea with milk may take the
place of one or two of the supplies of plain milk, say in the
morning and evening.   .

As natural sleep returns, and the patient's general condition improves, the stomach will become able to deal with
larger supplies ; therefore increase the amount by one-half,
and add to each supply some farinaceous material. Increase
also the interval to three hours, not awaking the patient to
take food, but getting in the supplies between his sleeps.
As convalescence becomes established the following diet will
be well borne :—

Early Breakfast in Bed.—A cup of tea or of milk with
an egg beaten up in it, a slice of toast or butter ; on alternate days a cup of cocoa with bread and butter.

In the Middle of the Forenoon.—A cup of good meat soup (Recipes 1–10, and 40), with toast, or a cup of warm milk and a biscuit.

Early Dinner, 1.30.—A small basin of plain *purée* of meat, with bread, or a *panada* of chicken with a little mashed potato, followed by a simple milk-pudding (Recipes 57–64). *Beverage*—A small glass of home-made lemonade or of milk and soda-water.

For Afternoon Tea.—A cup of cocoa or of tea, with plenty of milk, with a slice of bread and butter or a plain sponge-cake.

Supper, 7.30.—Fish or sweetbread or tripe, or a few oysters with brown bread and butter. *Beverage*—A glass of milk and soda-water.

In the later stages of convalescence the same diet as directed for the slighter forms of chronic alcoholism will be found suitable (p. 120).

Exclusion of Alcohol from the Dietary.—The advisability of excluding every form of alcohol from the dietary of patients suffering from delirium tremens is by some called in question, but it may safely be asserted that in the great majority of uncomplicated cases no harm follows the practice of cutting off alcoholic stimulants altogether. Where, however, there is great feebleness of constitution, especially in old people, or where some complication such as pneumonia sets in, the case is altered, and it may then be absolutely necessary to give some form of alcohol to prevent the patient from sinking. Spirits should even then be avoided, and if the patient can take it, some form of malt liquor given with food. In cases, too, where prolonged sleeplessness has become a source of anxiety to the medical attendant, it may be found necessary to give stout or other malt liquor if sedatives by themselves have failed to produce sleep.

Nutrient Enemata.—In all cases where sufficient nourishment cannot be taken by the mouth, or where the patients fight against their food, nutrient enemata should be freely given.

CHAPTER XV.

NERVOUS DISORDERS.

GENERAL CONTENTS : Hysteria—Weir-Mitchell Treatment.

HYSTERIA.

THE question of diet is one that should bulk very largely in the treatment of all hysterical patients. The food question is a constantly-recurring difficulty, and one that in some form or other meets the physician at every turn. Careful inquiry will usually prove that a sufficiency of food is not being taken, or that the diet is sadly deficient in some important particulars. Some hysterical patients like to pose as interesting invalids, who can eat next to nothing; others have so many likes and dislikes—the dislikes greatly preponderating—that the least digestible, least nourishing, and altogether most unsuitable articles of diet are the only things taken. In both classes a state of semi-starvation is kept up. The nervous system, already disordered, and requiring to be well nourished, is alternately excited and depressed by the unwholesome supplies of food and drink ; morbid ideas increase, and a vicious circle is established by these factors acting and re-acting on each other. The result is indigestion more or less constantly present.

Gradual Increase of Food.—It is useless in such circumstances to attempt to remedy this condition of affairs by putting the patient suddenly on to a full diet. We must begin by interdicting the improper foods. All stimulants must, save in exceptional cases, be absolutely forbidden; also strong tea and coffee. A dietary consisting largely of meat, to the exclusion of farinaceous substances, green vegetables, and fruit, is often indulged in by such patients, although it is not well adapted to their actual wants. What they do require is food in such forms as will readily supply nourishment to their disordered nervous systems—not an unlimited supply of nitrogenous material, for that would further tax their already weakened digestive organs. Eggs, fish, fat meats, butter, cream, milk, puddings made with milk, fresh vegetables, and cooked fruit should enter largely into the dietary we prescribe.

Small Meals.—Large quantities of food at a time are not well borne; hence four or five small meals will be the order of the day.

Breakfast.—A saucerful of porridge (well-boiled oatmeal, wheat-meal, or hominy) with cream, followed by a frizzle of fat bacon or an egg, and a slice of bread and butter, will form an excellent breakfast. In the middle of the forenoon a cup of milk with a biscuit.

Luncheon, at 1.30 or 2 o'clock.—A plain meal—two courses—fish or meat, and pudding; thus, a slice from the joint, a spoonful of mashed potato, and some well-boiled green vegetable. A milk-pudding or plain shape, with stewed fruit.

Beverage.—Half a tumblerful of cold water. In the afternoon a cup of weak tea, with milk, or a cup of cocoa, and a slice of toast or of bread and butter.

Supper, at 7 or 7.30.—A light supper, consisting of egg or fresh white fish, chicken, sweetbread, or the like, with bread and butter, and half a glass of water. Before going to bed a teacupful of milk (with the chill taken off), and a biscuit or slice of bread and butter.

If the patient wakes early in the morning, she should have within reach some milk, easily warmed by means of an Etna or other form of spirit-lamp, and in any case some

light food, such as a cup of cocoa or of warm milk should
be taken in the morning before dressing. Without this
precaution the patient is too much exhausted after the
bath and the fatigue of dressing, to have any appetite for
breakfast. With returning strength and the ability to
take larger quantities of food at regular meal-times, the
intermediate supplies of the forenoon and evening may be
gradually diminished and dispensed with.

Alcohol.—Allusion has already been made to the importance of excluding, in these cases, from the dietary alcohol
in every form. It must be granted, however, that where
the digestive functions have for a long time been seriously
impaired, and there is much consequent weakness, a limited
amount of alcohol will be necessary.

Light Wine.—The form in which it is least objectionable
is that of a light wine, and the patient may be allowed
to take with the midday meal and with supper a glass of
claret, of light burgundy, of hock, or of chablis. Experience proves that patients of this class who are able to
spend a considerable portion of time in the open air, and
to take a reasonable amount of exercise, do better on an
entirely non-alcoholic diet.

Slighter Cases.—In cases where things have not gone to
extremes, where friends and attendants have not altogether
lost control of the patient, judicious management, together
with properly regulated exercise and a diet such as that
sketched above, will often work wonders in the treatment
of hysterical patients.

"Weir-Mitchell" Treatment.—When, however, there is
great emaciation and serious depression of nerve power, it
is to what is generally known as the Weir-Mitchell plan
of treatment that we now turn in the hope of benefiting
these patients. The combined nature of this treatment
must not be forgotten—rest, isolation, massage, large quantities of food, and electricity. Without strict isolation and
complete rest, little good will be obtained, and without

massage properly carried out, the large quantities of food cannot be digested or assimilated.

Isolation.—By separation from friends and ordinary surroundings, old and often hurtful associations are broken up, and the worried nervous system is put at complete rest.

Massage.—The massage, with or without electricity, takes the place of exercise, and the amount can be regulated from time to time as the physician finds necessary to suit the requirements of the particular case. By means of systematic massage the elimination of waste products is greatly stimulated, and a steady building up of healthy tissue takes place, the ultimate result being a vast increase in both muscular and nerve power.

Selection of a Nurse.—The first important conditions of success are the selection of a suitable and reliable nurse, and of a bright airy room with a good exposure. These are matters of the greatest importance, when it is considered that the patient must remain in bed absolutely for at least three or four weeks from the beginning of the treatment, and that all communications with the outer world are practically cut off. Before commencing treatment, the patient's weight should be ascertained, and may be taken at intervals during the course.

Milk Alone at First.—At first the food given should be milk alone, with the exception, perhaps, of a small cup of black coffee early in the morning, if required to obviate a tendency to constipation. The milk will generally agree best if it be slightly warmed, but unless there be diarrhœa it need not be boiled. For the first day or two, give every two hours from three to four ounces of milk, which will bring the total quantity in the twenty-four hours to about two pints, allowing for the intervals of sleep.

Amount of Milk.—The amount given at a time should be gradually increased, and the interval lengthened to three hours, the total being brought up to two quarts. The milk should be sipped very slowly, and if it be very dis tasteful to the patient, as is the case in rare instances, it will be necessary to add enough tea, coffee, or cocoa just to flavour it. In other cases where milk is at first badly borne, though happily these also are rare, the difficulty may be got over by giving milk with a lessened amount of ordinary food, and gradually increasing the quantity of milk until it becomes the sole diet.

Alkalies with Milk.—Alkalies with milk will be needed only in those cases where it sets up much acidity. After two or three days alternate supplies of five and ten ounces every three hours will usually be well borne. Soon ten ounces every three hours can be taken with ease, and we shall next proceed to add for breakfast a cup of cocoa with bread and butter, and in the middle of the day a milk pudding. Fish or chicken will next be added, first either at the midday or evening meal and then at both. The diet will, therefore, stand thus :—

Milk.—Sixty to eighty ounces.

Breakfast.—Cocoa and bread and butter.

Luncheon.—Fish, vegetable, and pudding.

Dinner.—Chicken, vegetables, and bread.

In most cases, after a few days of this diet further additions may be made as follow, still keeping up the full quantity of milk.

Breakfast.—Porridge and cream early in the morning.

Second Breakfast.—Cocoa and egg, bread and butter.

Luncheon.—Fish, bread, pudding and milk ; or chicken, vegetables, and pudding.

Dinner.—Mutton or beef, two or three kinds of vegetables, milk-pudding, or stewed fruit with cream.

Extract of Malt may be given with one or more supplies of milk during the day, and if strong soup and beef peptonoids are used, they will take the place of one of the supplies of milk.

Cod-Liver-Oil and Iron are useful adjuncts in some cases, and other tonics, as acids, quinine, and strychnia, given towards the close of the treatment, may be usefully continued for some time afterwards.

Gradual Reduction of the Diet.—At the end of five or six weeks the slackening off of this excessive diet is commenced by reducing the quantity of milk, and as the patient comes to take more exercise, the massage also is lessened.

Increase in weight and muscular power are the best tests of the efficiency of the massage. If the result be not satisfactory in these two important respects, it is high time to inquire closely into the way in which the massage is being carried out.

After Treatment.—When the term of treatment is concluded, the patient should lead an active open-air life, but for some months longer a daily rest of two or three hours should be inculcated. The return to ordinary diet must be made gradually, by leaving off the intermediate supplies of food one by one. In this way the system by degrees adapts itself to the change. A sea-voyage or a prolonged tour in many cases confirms the cure. Relapses happily are uncommon, especially where this treatment is followed by a life of active usefulness.

CHAPTER XVI.

DIATHETIC DISEASES.

FEVERS (EXCEPT TYPHOID FEVER, *see p.* 142)—ACUTE PNEUMONIA, &C.

THE teaching of Graves marks a revolution in the dietetic treatment of fevers. A great reaction then set in against starving in acute diseases accompanied by fever, and since that time the danger has been more on the side of over-feeding than of under-feeding.

Precise Directions.—Unless precise directions are given by the physician, anxious friends, and even nurses, who should know better, are apt to push the quantity of food too far, thereby oftentimes adding to the discomfort of the patient, if they do not actually endanger his life or retard his recovery.

A very prevalent idea is that if the patient be restricted to beef-tea, broths, and milk, these foods may be given practically without limitation, to make up for increased metabolism and loss of nitrogenous materials consequent on the feverish state. Undoubtedly the difficulty in not a few cases is to get the patient to take a sufficiency of nourishment; but, on the other hand, the unlimited administration of food in cases where the patient will take it is as reprehensible as the plan of starvation.

Several general points should here be noted in regard to

137

the febrile state and the feeding of patients during the course of acute febrile diseases.

Impairment of Digestion.—In the first place, we find that the functional activity of the digestive organs is seriously impaired from an early period of the febrile attack. Sir William Roberts has stated forcibly (*Dietetics and Dyspepsia,* p. 67) the distinction between gastric and intestinal diges-tion, pointing out that in persons who are seriously ill the stomach is inactive, and the secretion of gastric juice so greatly diminished as hardly to be taken into account in fixing the diet. He believes that in severe cases the stomach loses its normal office, and becomes merely a con-tinuation, as it were, of the œsophagus to carry on the liquid food to the duodenum. He remarks that this loss of gastric activity accounts for the fact that many persons who in health cannot digest milk are able to do so with ease during illness. There is not enough gastric juice to curdle the milk, so it passes on into the bowel unchanged, and is there readily acted upon by the pancreatic secretion.

Curds in the Stools.—The passage of undigested food and of hard masses of milk-curds in the motions is an indication of some error in the dietary that should never be overlooked. When this takes place, food is either being given of such a kind or in such quantity as the enfeebled digestive organs cannot cope with. If the quantity or quality be not at fault, then we should suspect the mode of administration of the food or the circumstances under which it is given. Food that is not digested and assimilated, besides causing discomfort to the patient and producing an increase of fever, may also (as, for example, in typhoid fever) be actually an additional source of danger to life.

Fluid Foods only to be Given.—It follows, therefore, that only the simplest kinds of fluid food should be given, as will be fully pointed out when discussing the articles of the diet.

Secondly, throughout the course of a febrile attack there

is greatly increased waste of tissue, proportionate to the rise
of temperature and the continuance of the fever. The nitrogenous tissues suffer most, but also secondarily fats, as shown
by wasting of muscles and by general emaciation.

Limitation of Albuminoids.—The rapid destruction of the
nitrogenous tissues also leads to an accumulation of nitrogenous waste materials in the blood, and in all cases where
the elimination is to any extent defective, as it is in most
fevers, the quantity of albuminoid substances in the diet
must be kept down to a small proportion.

It is very doubtful to what extent albuminoid foods can
be made use of by the organism during the continuance of
fever in the formation of new tissue to replace that lost by
increased metamorphosis, but the judicious selection and
combination of foods in the dietary may prevent undue
waste, and afford the patient sufficient material for organic
combustion, or, as it is sometimes called, a sufficiency of
"fuel food."

Milk.—The staple food in febrile conditions is undoubtedly
milk, but caution is necessary in the use of even milk, for
if it be given without care and discrimination, untoward
results may follow. As has been already noticed, milk may
pass along the intestinal canal, remaining almost in its fluid
condition during febrile states; but, on the other hand, if
milk be given without due precaution, firm curds will be left
undigested, and these, especially in cases where the bowels
are inflamed or ulcerated, may do serious mischief.

Diluted Milk.—To prevent excessive curdling of milk, it
must be given in small quantities at a time, say from two to
six or eight ounces, diluted with lime-water, or with soda,
seltzer, or other effervescent water—one part of such water
to two of milk. Where it is desirable to limit the bulk,
saccharated solution of lime, ℥xx. to each feederful of milk,
may be used, but generally a considerable quantity of fluid
is needed to supply the loss of moisture that is taking place
in the febrile state, and hence the other methods, namely,

by dilution, are to be preferred. If milk is well borne, from two to three pints should be given—four ounces every two hours or six ounces every three hours, which will come to about two and a half pints in the twenty-four hours. The quantities, the hours at which they are given, and the total amount, should be accurately put down by the nurse.

Whey.—In cases where milk does not agree well, whey (Recipes 80, 81) may be given as a substitute for milk in part or in whole. Sometimes the milk is taken more easily and readily if we permit certain of the supplies to be flavoured with tea, coffee, or cocoa, and there is usually no serious objection to doing so.

Beef-Tea and Broths.—Next to milk, the most important articles of diet in fevers are beef-tea, meat broths, and chicken-tea, and these are much more valuable as foods when they contain some farinaceous substance, such as baked flour or pounded biscuit. Farinaceous substances are required, and are useful by affording fuel food. Arrow-root made very thin, gruel also thin and carefully strained, are types of this form.

Prepared Foods.—But there is here a large field for the use of prepared foods, of which Mellin's and Benger's may be taken as typical examples. These foods readily supply the grape-sugar that is more than ever needed to meet the loss occasioned by rapid oxidation during the continuance of high temperatures. Malt extracts are also valuable additions to the dietary.

Thirst.—Usually there is considerable thirst in fevers, and patients take readily of diluent drinks. Even if not asked for, simple drink should be given between the supplies of food, and they may not only be varied, but may have some nutritive value by the addition of sugar or of malt extracts.

Pure cold water is often the most acceptable drink to the patient, but barley-water or rice-water, and beverages made with the juice of fresh fruits, such as currants (Recipe 20), and lemonade (Recipes 24, 25), are useful alternatives; also the old-fashioned potus imperialis (Recipe 21), and toast-water.

Alcohol in Fevers.—Alcohol is often, but by no means always, a necessary adjunct to the diet in fevers and acute diseases generally. It should not be prescribed in all cases as a matter of course and as a part of the recognised dietary. In acute diseases lasting but a limited time, in an ordinarily healthy person, alcohol is not required at all, or not at least until the convalescing stage, when a little wine may improve the appetite and aid digestion.

Indications for the Use of Alcohol.—In all cases, however, when any sign of failure of circulation shows itself; in those cases in which severe symptoms and high temperatures appear early in the course of the disease; in persons who have been accustomed to the use of large quantities of stimulants; in the aged, in very young children, and in all persons of feeble constitution, alcohol is required, and should be given freely, if necessary, even from the commencement of the attack. Whenever and wherever we have to face the signs of feeble circulation and of nervous prostration in acute disease, alcohol is necessary. The small compressible, rapid pulse, the feeble or scarcely audible first sound of the heart, with or without low muttering delirium, are sure indications for the use of alcoholic stimulants.

To get the full good of alcohol, in whatever form it may be used, it should be given with or just after food, and thus too great borrowing of strength is prevented.

Spirits.—(a.) During the course of febrile diseases, the best form of alcoholic stimulant is good old spirit, and, as a rule, brandy or whisky answers best. The quantity will vary with the age and condition of the patient. The minimum, however, may be put at two ounces, and the average at from three to six ounces. In exceptional cases, however, as much as twelve ounces or more may be required.

Wines.—(b.) If there be any reason to prefer giving wine

rather than spirit, then a good light port, or burgundy, or a champagne may be selected. It is during convalescence, however, as already noticed, that wines are mostly useful, and the rule as to quantity should be, that no more be given than that amount which increases appetite and improves digestion.

Typhoid Fever.

The regulation of the diet in typhoid (enteric) fever forms such an important part of the treatment, that it is desirable to consider it by itself, apart from other fevers, and at some length.

Danger of Over-Feeding.—The risk of over-feeding has already been alluded to in speaking of the diet of fevers in general, and it need only here be remarked, that the tendency to over-feeding is perhaps greater in typhoid than in other fevers and febrile diseases, whilst the danger to the patient is likewise greater.

Condition of the Stools.—The chief guide in the matter of food in typhoid fever must be found in the condition of the stools. The physician in charge of the case should himself see the motions daily, and should not take this important matter on the report of even an experienced nurse, since she may easily be misled by appearances. If any signs of undigested food show themselves, there is something regarding the dietary needing to be inquired into, and, if possible, put right.

Milk.—A very common idea is, that a typhoid fever patient cannot take too much milk. This is probably true, up to the point to which the milk is fully digested, but no further. It is undigested food, and especially undigested milk in hard curds, that causes the danger in the inflamed and ulcerated state of the bowels. If masses of hard curd appear in the motions, there is some fault in the dietary. Too much milk is being given in the twenty-four hours, or it is being given in too large quantity at a.time, and these points must be looked into and carefully considered.

Alkali with Milk.—(*a.*) It may be that the quantities are not wrong, and that dilution of the milk, or the addition of an alkali, or of some farinaceous material in powder, will answer the purpose, by preventing the formation of firm curds, which pass along undigested, and add so greatly to the risk of increased diarrhœa, of hæmorrhage, or even of perforation. Remembering the condition of the bowels, it cannot be wondered that the passage of hard masses of curdled milk should cause such discomfort to the patient in the way of flatulence, griping, and diarrhœa, if worse results even do not follow.

If the total quantity of milk given be not too great, and yet it be not fully digested, careful inquiry must be made into the mode of administration of the milk. The quantity given at a time should be restricted, or if the fault do not lie there, an alkali such as lime-water or soda-water should be added to the milk. Where it is desirable not to increase the bulk too much, and not to give it quite cold, the chill may be taken off by adding a little hot water, while the addition of twenty drops of the saccharated solution of lime in each supply will secure alkalinity.

Farinaceous Substances in Milk.—(*b.*) Another useful measure is to put some farinaceous substance, such as arrowroot or baked flour, into the milk, to aid the subdivision of the curds. Such a substance as baked flour, containing as it does a proportion of dextrine, forms a considerable addition to the nutritive value of the food.

Precise Directions.—As has been already pointed out, definite directions should be laid down as to the quantity of food to be given at a time, the mode of its administration, the intervals between the supplies, and the total amount for the twenty-four hours. For example, it should not be left to the nurse to give, if she thinks fit, milk and water as an ordinary drink whenever the patient is thirsty.

He may, however, be allowed in the intervals between the
supplies of food to sip freely of plain cold water, or of
toast and water, barley-water, lemon-juice in water, or other
such simple drinks (Recipes 19, 22, 24, 31, 32).

If milk-sugar or grape-sugar in some form be added, it
helps materially in sustaining the patient.

Beef-Tea and Broths.—Hitherto mention has been made
of milk only, but beef-tea, properly made (Recipe 3),
mutton-broth, chicken-tea thickened with farinaceous substances, may be given in greater or less quantity, according
to circumstances.   One point that should regulate the
amount of meat-broths given is the state of the bowels.
If diarrhœa be at all an urgent symptom, then beef-tea
and broths are contra-indicated, or should be given in very
small quantities.   In any case of typhoid fever, meat-
broths, if given in large amount, tend to increase diarrhœa.
In some cases meat-jelly iced or extracts of meat, may
take the place of beef-tea and broth, a teaspoonful being
given at a time.

Dietaries.—Coming to the details of the diet, in an ordinary case two and a half pints of milk and a pint and a half
of beef-tea or of mutton or chicken broth, will be a fair
average supply, given in divided quantities, alternating as
far as possible the milk and the broth.   In this way, giving
ten ounces at a time every three hours, the broth will come
in after about every second supply of milk.   Where as much
as ten ounces cannot be taken comfortably at one time and
well digested, we may give from five to seven ounces every
two hours.

Sleep.—An important question is whether the patient
should or should not be wakened to take nourishment.   The
judgment of a sensible, observant nurse may in this matter
be trusted to a considerable extent, but the patient should
certainly not be allowed to miss more than one supply of
food; and often less harm will be done by gently interrupt-
ing a somewhat restless sleep to give food than by risking

the exhaustion consequent upon the patient going too long without nourishment.

Nutrient Enemata.—Nutrient enemata are sometimes required to maintain life and tide over dangerous periods in typhoid fever (for particulars see chap. on Nutrient Enemata). If they are not well retained, from fifteen to twenty-five drops of laudanum should be added to each enema.

When serious hæmorrhage occurs, it is necessary to suspend feeding by the mouth altogether for twelve or twenty-four hours, and to give small enemata of prepared and predigested materials, with or without stimulants according to the condition of the patient. If he is very weak and collapsed, a moderate amount of alcohol must be given, but not more than is absolutely necessary, lest by this means the tendency to hæmorrhage be increased.

Alcohol.—Next to the consideration of the food comes the question of alcoholic stimulants in cases of typhoid fever.

A considerable number of cases can no doubt be treated successfully without the use of alcohol in any form, and it may almost be said that it is never required in the early stages of the disease. The exceptions to this rule are patients in a very debilitated state, and those who have previously been accustomed to imbibe freely; also cases in which very high temperatures and other serious symptoms come on suddenly at an early period. In such cases it is necessary to give alcohol without delay, perhaps even from the beginning; but in ordinary cases it should never be ordered as a matter of course or without due consideration of the circumstances of the individual patient.

Indications for the Use of Alcohol.—Our chief guide to the use of·alcohol in typhoid fever lies in the condition of the heart. A small, frequent, easily-compressed pulse, especially if associated with feebleness of the first sound of the heart, is a clear indication that alcohol is required. If under its use the pulse becomes steady, fuller, less frequent, and less easily compressed, the alcohol is doing good. When along

K

with this improvement of circulation there is lowering of the temperature and less delirium, there remains no doubt as to its beneficial effect.

Quantity of Alcohol to be Given.—The amount given at a time should be small—from a teaspoonful to a tablespoonful of spirit, or an ounce of wine three or four times a day. The stimulant should be given in the food or just after it, and the effect must be carefully watched.

If required, the quantity given in twenty-four hours may be increased up to eight or ten ounces, in divided doses of half an ounce or an ounce each. In cases of extreme weakness, an ounce of spirit every hour will not be too much ; and it is astonishing how well patients who in health are unaccustomed to the use of stimulants bear so large a quantity of alcohol. Good old port or madeira answers well; but, on the whole, preference is to be given to mature spirit in the form of old brandy or whisky.

In cases where the cold bath or cold sponging has to be resorted to, it is desirable, as a rule, to give from two teaspoonfuls to a tablespoonful of spirit in a little water just before commencing or during the process.

Diet in Convalescence.—During convalescence the regulation of the diet is of the utmost importance. The condition of the bowel demands the greatest care on our part that the tender healing, or newly healed, surfaces be not injured by any hard or indigestible substance. The worst consequences have followed very slight negligence in this respect, and, therefore, for a considerable time the greatest caution is necessary. It is a safe rule to follow, that no solid food be given until the temperature has been normal for at least eight days, and in irregular and severe cases a longer period should elapse. Even then, it is not uncommon to have a slight rise of temperature with some constitutional disturbance on the first return to really solid food.

The first step should be to increase the thickness of the beef-tea or soup with fine bread-crumb. Next, lightly-made farinaceous puddings may be given, also custards and jellies. Later on, beaten-up eggs and slightly-boiled and poached eggs may be allowed, and also finely-pounded meat. As a further step, tiny sandwiches made of pounded chicken or pheasant, between thin squares of bread, will often be greatly relished by the patient. The way for red meats should be prepared by the giving of a slice from the breast of a chicken with a little bread-sauce. Oysters also form an agreeable change. White fish is useful and allowable early; but the danger of bones being left in it, even by the most painstaking cook, makes us chary of its use.

Fruits.—Fruit is contra-indicated throughout the course of typhoid fever on account of the risk there always is that it will increase diarrhœa, but in the convalescent stages a few grapes may be allowed, provided that the skins and pips are removed and not swallowed. The pulp of a ripe orange may be admitted under similar stringent conditions.

If the patient has been taking any considerable quantity of alcohol during the acute stages of the disease, the amount may be gradually lessened during convalescence, and the form of the stimulant changed to light wines given with the food. Green vegetables and fruits, other than those mentioned above, must be avoided for a considerable time, and the return to ordinary habits of diet and of life must be made very carefully and very gradually.

Note.—It is best to have separate feeding cups for food and medicines. If the same cup be used, or even one that is similar in appearance, the association of ideas is against the taking of food.

Cleansing of the Mouth.—It is hardly necessary to remark here that frequent cleansing of the mouth tends to make the taking of food easier. When the mouth is allowed to become dry and foul, this condition is a serious obstacle to the taking of nourishment.

CHAPTER XVII.

DISEASES OF CHILDREN.

INDIGESTION IN INFANTS AND IN OLDER CHILDREN.

(*a.*) **Healthy Milk—In Infants.**—If the child is being brought up by hand, great care must be taken, if symptoms of indigestion appear, to see that the cow's milk is perfectly healthy. If no fault is to be found with the milk, all that may be needed is for a time to dilute the milk with a larger proportion of water than usual. The diet mentioned in the section on Diarrhœa will also often agree well in such cases, or cream diluted with four or five times as much water, or the following diet, recommended by Drs. Meigs and Pepper for weakly children, consisting of prepared gelatine or isinglass, cow's milk, cream, and a very thin arrowroot-water with sugar (p. 153). They advise as an alternative to this diet, one composed of thin arrowroot-water, lime-water, cream, and milk in equal proportions; the quantity given at first to be not more than four tablespoonfuls every two hours. After a day or two, if the food is well borne, six tablespoonfuls are to be given every two hours, and this again increased to eight tablespoonfuls. As the gastric disturbance subsides, the quantities of cream and lime-water are to be lessened, and the amount of milk increased, thus bringing the child gradually back to its ordinary diet. During acute attacks the child should be allowed to drink freely, in the intervals between the supplies of food, of fresh cold water. When there is great prostration, the water should contain brandy, in the proportion of two teaspoonfuls to the pint.

Indigestion without Diarrhœa.—When no diarrhœa is present, weak mutton- or chicken-broths will often suit when other things fail. Strict regularity in the times of giving the food, and careful regulation of the quantities given at one time, are of the utmost importance.

(b.) In Older Children.—Restriction of the diet to bread and milk, milk-puddings, bread and butter, broths thickened with some farinaceous substance, such as baked flour, rice, or crumbled toast, pounded chicken, rabbit, or very tender mutton pounded, with little or no vegetable for the time being, and the exclusion of all sweets, cakes, and fruits, will answer best, and that often without the aid of drugs.

Dietary.—In detail such a diet will be :—

Breakfast.—Very well-boiled whole meal or oatmeal porridge with milk, or bread and milk, or cocoa and bread and butter.

Dinner.—Chicken or tender mutton (pounded, if necessary), with bread and a little dry boiled rice, or broth thickened, followed by a little milky pudding.

Tea.—Warmed milk, or milk and hot-water, or cocoa, with bread and butter.

Supper.—A little milk and a plain biscuit.

Children should be made to eat very slowly, and to masticate food very thoroughly.

CHRONIC DIARRHŒA.

The food composed of arrowroot, cow's milk, and cream with gelatine is an exceedingly useful one.

Dr. Eustace Smith recommends, for a child one year old, who can digest a certain quantity of milk, and in whom diarrhœa is not very severe, a teaspoonful of Liebig's food for infants, every three hours, dissolved alternately in equal parts of milk and water, and in equal parts of weak veal-broth and barley-water.

If the child can digest no milk, he advises the following diet :—

7 A.M.—One teaspoonful of Mellin's food dissolved in a teacupful of veal-broth and barley-water, equal parts.

11 A.M.—One tablespoonful of cream in a teacupful of fresh whey.

2 *o'clock.*—The unboiled yolk of one egg beaten up with fifteen drops of brandy, a tablespoonful of cinnamon-water and a little white sugar.

5 P.M.—Six ounces of beef-tea (1 lb. to the pint).

11 P.M.—A meal similar to the one at 7 A.M.

As the child's condition improves, a little milk can be gradually added to the diet.

RICKETS.

While different theories regarding the pathology of rickets may be maintained, authorities on the subject are agreed as to the causation of the disease.

Causes.—Defective nutrition of children is without doubt the usual cause of rickets, and the dyspepsias and gastro-intestinal catarrhs due to improper feeding greatly favour the occurrence of the disease. Added to unsuitable diet, there are usually, though by no means invariably, to be found bad hygienic conditions—a low damp situation, bad ventilation, deficiency of exercise in the open air, and want of cleanliness.

"Rickets, although one of the most preventible of children's diseases, is one of the most common. It begins insidiously, presenting at first merely the ordinary symptoms of defective assimilation, and attention is often not attracted to it until the characteristic changes occur in the bones which place the existence of the disease beyond a doubt. It is the result of mal-nutrition; any disease, therefore, which seriously interferes with the assimilative power, and causes sufficient impairment of the general strength, may be followed directly by the disorder under consideration, without any intervening stage. Reduce the strength to a given point, and rickets begins. Prolong this state of debility sufficiently, and the characteristic changes resulting from the disease manifest themselves. Any case, therefore, which will reduce the strength to this point, lays the foundation of rickets."

" Rickets is usually ranked amongst the diathetic diseases of childhood, but its claims to such a position are by no means indisputable. In rickets there is, strictly speaking, no constitutional predisposition; it is the result of certain known causes, without which the disease cannot be produced. That the disease occurs amongst the children of the rich as well as amongst the poor, is no argument against this view; for wealth cannot buy judgment, and education is no guarantee against foolish indulgence. We know that a child may be in reality starving, although fed every day upon the richest food, for he is nourished, not in proportion to the nutritive properties of the food he swallows, but in proportion to his capability of digesting what is given to him. . . . Rickets does not produce mal-nutrition; mal-nutrition produces rickets."

" Rickets, then, is not a diathetic disease in the sense in which tuberculosis and syphilis are diathetic diseases. It is acquired under the influence of certain causes, lasts as long as these causes continue in operation, and, unless the structural changes are so extensive, and the general strength so reduced as to forbid recovery, passes off when the causes are removed."—*Eustace Smith.*

Allusion has already been made to the part that bad hygienic surroundings take in the causation of this disease, and acute diseases which rapidly exhaust the strength of the child and lower his powers of assimilation, should not be omitted from the list of common causes.

Health of the Mother.—If the mother herself be in feeble health whilst she is suckling her child, her milk will not be sufficiently nourishing, and the sooner the child is transferred to a wet-nurse or weaned, the better. Even if the mother be in good health and be well-fed, her milk becomes, after a certain time, thin, watery, and insufficient for the wants of the growing child; hence the evil effects of prolonging lactation beyond the recognised period of nine months. It is usually the case that the milk of a healthy mother entirely suffices for the wants of the child during the first six months; and may be continued as its staple food for nine months, but during the latter portion of that time, a little carefully-selected and well-prepared farinaceous food should be added to the dietary. When, from any cause, the mother's milk proves insufficient for the needs of a young infant, a healthy wet-nurse should be obtained, or the child fed upon cow's milk or ass's milk, if that be procurable. If fed upon the mother's milk alone, the infant's nutrition will assuredly in such circumstances suffer, and it will begin to show signs of feeble health, which, if not arrested, may lead up to the development of rickets.

Excess of Farinaceous Foods.—Too often the attempt is made to supply the deficiency in the milk by giving to young children farinaceous foods in quantities which they cannot digest, and which, therefore, since they produce indigestion and flatulence, only increase the mischief.

Drs. Meigs and Pepper, in their work on the diseases of children, while they deprecate most strongly the giving of a large amount of farinaceous food, yet consider that a small amount of starchy material given in the milk, by preventing its coagulation into hard masses, does sometimes make the food more digestible. For young infants they give the preference to arrowroot. Where there is a tendency to diarrhœa, they recommend barley or wheat; and when the bowels are constipated, oatmeal. They have long used the following preparation, and believe it to be one of the best substitutes for the natural food:—"To make this food, a scruple by weight of Russian isinglass or prepared gelatine, or a portion of gelatine cake two inches square, is soaked for a short time in half a pint of cold water. The water is then boiled until the gelatine is fully dissolved—about fifteen minutes. A small teaspoonful of arrowroot mixed into a paste with a little water is then stirred into the boiling water, after which the milk is added and allowed to boil for a few minutes. At the end of the boiling the cream is added."

Proportions of Milk.—The proportions of milk are—for the youngest children, one-third; and for the older, one-half or two-thirds. Of cream, two tablespoonfuls are added to a pint of the food, so long as this is one-third milk. When the food is half milk, one tablespoonful and a half of cream to the pint is the proper quantity; and when the food becomes two-thirds milk, one tablespoonful is to be added. To dilutions of two-thirds water and one-third milk there should be added about six and a half drachms of the sugar of milk; to dilutions of half-and-half, the quantity of sugar to be added is five and a half to the pint. If cane-sugar is used, only half the above quantities should be employed. They further recommend that in the case of very weakly children the food should be more diluted than the proportions given above. For older children the food may be somewhat strengthened.

Dietary.—For a child sixteen or eighteen months old, suffering from rickets, Dr. Eustace Smith recommends the following diet:—

7.30 A.M.—One or two teaspoonfuls of Liebig's food for infants (Mellin's) dissolved in a breakfast-cupful of milk.

11 A.M.—A breakfast-cupful of milk, with fifteen drops of the saccharated solution of lime in it.

2 P.M.—A good tablespoonful of well-pounded mutton chop, with gravy, and a little crumbled stale bread; or a

good tablespoonful of the flower of broccoli, well stewed with gravy until quite tender; a little dry bread. For drink, milk and water.

6 P.M.—Same as the first meal, or, if no meat has been given, the lightly-boiled yolk of one egg; a little thin bread and butter. Milk and water.

Later on a considerable proportion of animal food is desirable, and where complete mastication cannot be secured, the meat must be pounded. Tonics, with iron, are frequently useful, but most reliance should be placed on cod-liver oil in addition to a liberal dietary. If difficulty is experienced in getting the oil well taken, a useful formula is the following, given by Drs. Meigs and Pepper:—

℞.

Ol. Jec. A-elli .	.	.	.	f.ʒj.
Pulv. Acaciæ .	.	.	.	q.s.
Ol. Cinnamomi	.	.	.	gtt. vj.
Sacch. Alb. .	.	.	.	q. s.
Aq. Cinnamomi	.	.	.	q.s. ad f.ʒiij.

Dose.—A dessertspoonful three times a day after eating.

NEUROSAL AFFECTIONS OF CHILDREN.

In these days of competition and competitive examinations, when to most people success in life seems to depend not so much on good bodily growth as on early mental attainments, there is serious danger that in the case of children physical development may be neglected.

Over-Pressure.—The mischief is often to a great extent done before medical advice is sought. The doctor is consulted only when the child has become pale, thin, and restless; has lost appetite, starts and talks in sleep, and above all, has lost interest in lessons. The other symptoms may, perhaps, go unheeded, but this one soon attracts attention. On inquiry, it will be found in the majority of such cases, that the child's diet is thoroughly unsuited to the wants of the system ; that its exercise and open-air life are limited, and its lessons heavy.   No wonder, then, that nutrition flags and the child falls out of health.   We are concerned here with the question of diet only ; we must point out wherein the error lies, and what is the remedy therefor.

Unsuitable Diet.—To feed young, delicate children, of an excitable and unstable nervous system, on a stimulating dietary, consisting largely of animal food, and to give them tea and coffee, and any form of alcoholic stimulant, with a view to strengthening them, is to overlook altogether and disregard the relation of special foods to particular constitutions and states of system.   A highly nitrogenous diet, especially in cases where the child does not take much active exercise in the open air, besides worrying the nervous system, with excess of waste products, throws a great strain on the excretory organs.   The urine contains a deposit of lithates ; the bowels become constipated or irregular in action—constipation alternating with diarrhœa—and this condition of mal-assimilation and defective nutrition often lays the foundation for serious organic disease in after-life.

Defective Nutrition.—This defective nutrition, in these cases from over-feeding of a kind, is just as marked and as serious to deal with as the same condition, produced by an insufficient amount of food, in the children of the poor. Another way in which impaired digestive power is produced and kept up, is by the evil habit, so often contracted by children, of eating cake, buns, or sweets between meals. The appetite for plain substantial food is taken away, and the stomach is unable to do its work properly at regular meal-times.   Hence the twofold evil—an insufficient quantity of good food, and imperfect digestion of the food that is taken.

Plain Food.—Healthy children, who have plenty of open-air exercise, do not require tempting dishes to make them eat, and, on the other hand, provided the child has been trained to eat its food slowly, and to masticate it thoroughly, there is little fear of a healthy child eating too much plain food.

" The nutritive functions are so predominantly active for the purpose of carrying on growth and repairing the rapid waste caused by youthful activity, that if the natural craving for exercise in the open air be freely indulged, and due attention be given to the development of the bodily frame, the young may very safely be left to choose for themselves both the quality and quantity of their food.   In such circumstances the natural taste inclines so essentially to the preference of plain substantial nourishment, that there is very little risk of excess being committed.   But where the parents are intent only on the intellectual advancement of their children, and accustomed to subject them daily to many successive hours of confinement and study, with only an hour or so of relaxation in the open air, as is too commonly the case both with those educated at home and in boarding-schools, an artificial state of being is induced, which makes the rule no longer applicable, and renders necessary a more careful attention to dietetic regimen."

Animal Food.—" Among the higher classes of society, the unrestricted use of the most exciting kinds and preparations of animal food, and the daily use of wine, are the means generally resorted to for the removal of the delicacy thus engendered; but when we consider the real state of the case, no remedy can seem more preposterous.   The evils to be corrected are, imperfect nutrition and want of strength. Now the imperfect nutrition is caused, not by deficient food, but by impaired powers of digestion and assimilation ; and these suffer because the lungs are denied the free air, the muscles their necessary exercise, the brain its cheerful recreation, and the circulation the healthy stimulus which these united conditions infallibly produce."—*Combe.*

Neurotic Children.—The children of neurotic parents, inheriting as they do, often in an increased degree, the parental tendencies, suffer most under bad dietetic arrangements, and frequently develop symptoms of definite nerve disorder. Keen at lessons, highly sensitive, disinclined for exercise and boisterous games, they are often to be found in a quiet corner with a book when other more robust children are romping and shouting in the open air. They eat poorly of simple food, and often have a craving for stimulating and unwholesome articles of diet. These are for the most part the children that come under our notice, presenting the symptoms we have briefly enumerated, and they require careful dieting. The notion that all children should be brought up alike is a too common one; and yet there is good reason for believing that much permanent good or harm does result from the continuance of a suitable or an unsuitable dietary in childhood and during the period of rapid growth.

Delicate Children.—We have in passing alluded to the diet of healthy children, placed in favourable hygienic conditions, and have seen that no special care beyond the selection of plain wholesome food is required; but it is different in regard to delicate and neurotic children, with whose case we are now more particularly concerned. Their food must be more carefully selected, must be more dainty, and must be given in smaller quantities at a time, until the digestive organs have regained vigour, when a return to ordinary diet will be permissible. Children of this type have little appetite in the morning. They should have no work to do before breakfast, and they should not be up and about so early as to be exhausted before the breakfast-hour arrives.

Dietary.—*Breakfast.*—If a small plate of porridge and cream, followed by a slice of brown bread and butter, can be taken, it forms an excellent meal. Some children prefer bread and boiled-milk, or rusks and milk, now and then for the sake of variety, which is often a point of importance in these cases; a cup of cocoa with plenty of milk in it, brown bread and butter, with an egg or a slice of bacon, may be substituted.

Lunch.—In the middle of the forenoon a teacupful of milk and a biscuit, or bread and butter.

The *Midday Dinner* should be the chief meal of the day, and should consist of a small slice of tenderly cooked meat or chicken, or a piece of fish with a little mashed or grated potato and well-boiled green vegetables, and a milk-pudding with stewed fruit; a small glass of water may be given, but not till towards the close of the meal.

Tea should be a small meal, if the dinner has been adequate—a cup of warmed milk or of cocoa, with brown bread and butter. If dinner has not been sufficiently hearty, an egg may be given at tea.

Supper.—A cup of bread and milk, or a plain milk-pudding.

Fruit.— Ripe fruit may be added at breakfast, and is especially useful where there is a tendency to constipation. In some cases of greater delicacy it may be advisable to give the egg beaten up in milk at breakfast or at tea.

Fat.—It is most desirable to have a good proportion of fat in some form in the dietary. Where it cannot be taken in sufficient quantity, as fat of bacon or of beef, or as butter and cream, it should be given in the shape of cod-liver oil.

CHAPTER XVIII.

ALCOHOL.

ALCOHOL.

A SHORT consideration of the place and uses of alcohol can hardly be separated from questions of clinical dietetics.

However strong the feeling may be that the regular daily use of alcoholic beverages is unnecessary, or even hurtful, to the majority of healthy people, it can hardly be denied by the most ardent advocates of total abstinence that in many cases of illness alcohol is not only a beneficial, but a necessary adjunct to the treatment.

The action of alcohol upon the body may be divided into its local and its general effects, and the latter subdivided into immediate effects and those that are remote.

Effects of Small Doses.—Suppose a person totally unaccustomed to the use of alcohol to sip for the first time a glass of, say, sherry or of port wine. He will describe the sensation in the mouth as one of some astringency, with slight burning or tingling, and with an increase in the flow of saliva. After the wine has reached the stomach, the circulation through the gastric mucous membrane becomes more active, and there is an increased secretion of gastric juice, with a feeling of warmth and a keener appetite for food. Next it will be noticed that the heart's action is stronger, and the pulse increased in force and in frequency.

The circulation through the different organs of the body is fuller, and dilatation of the capillaries of the skin is evidenced by flushing of the face. The subject of the experiment will declare that he feels warmer, but the actual temperature of the body is found to be slightly lowered, owing to increased perspiration, and the fact that the dilatation of the vessels of the skin brings more blood to the surface, thus in a cool atmosphere more rapidly reducing the temperature of the blood.

The effect upon the nervous system is one of stimulation, all the nervous centres taking on increased activity, and the higher centres being affected before the lower ones. The feelings and emotions are excited, the imagination is more lively, and there is, for the time being, a feeling of greater physical strength. Our friend, after his glass of wine, is at first slightly talkative, perhaps argumentative; presently he becomes more affectionate, and, if the dose be powerful enough, there follows thickness of utterance, disinclination for mental effort, and a tendency to drowsiness.

Effects of Large Doses.—As regards the digestive organs, large quantities of alcohol pass beyond the point of stimulation and act as irritants, thus retarding the circulation, and hampering the digestive processes (*cf.* p. 9).

Large amounts of alcohol produce effects upon the circulation and nervous system similar to those induced by small quantities, and in the same order, but in greater degree. The stage of excitement is more marked, and is rapidly followed by reaction and depression of function.

Recurring to the example already taken, if such a man as we have supposed imbibes a somewhat larger quantity of alcohol, or if any one accustomed to its habitual use takes a much larger amount than ordinary, all the symptoms of excitement are markedly present, but he speedily becomes intellectually dull and ceases to talk, the senses and finer feelings are blunted, then the voluntary muscles are affected, and he walks with difficulty or cannot walk at all—cannot even stand. His breathing becomes slow and laboured ; beginning at the extremities, the whole body becomes cold, and death may result from paralysis of respiration, or from failure of the heart.

These facts explain why deaths from exposure readily occur in persons under the influence of alcohol, and why alcoholic beverages are unsuitable for use by those who are continuously exposed in cold climates to very low temperatures.

Alcohol a Food.—It has often been questioned whether in reality alcohol is a food, but the balance of evidence is undoubtedly on the affirmative side. Alcohol, properly used, is a food of considerable value ; and on the other hand, when improperly used, and when taken in excess, alcohol is a poison.

In health, when plenty of food is available and food is well assimilated, alcohol is not required; but when food is scanty and assimilation defective, alcohol given in small quantity with the food not only aids digestion and assimilation, but, by to some extent retarding oxidation, it makes the food go further than it would otherwise do. "Hammond found that when, on insufficient diet, he was losing weight, the addition of a little alcohol not only enabled him to reach his former weight, but to add to it" (*Brunton*).

Experimental Evidence.—The recent very interesting and valuable experiments of Sir William Roberts show what effects are produced upon both salivary digestion by small and by large quantities of alcohol. He found that spirits, in not greater proportion than about five per cent., did not interfere with salivary digestion, but rather tended to aid it by increasing the flow of saliva. The action of wines he found decidedly inhibitory, due to their acidity ; for when the acid was neutralised, there was no retardation. The same statement applies to beer.

As regards gastric digestion, different kinds of spirit were found very much alike in their action, which depended on the amount of alcohol they contained. Up to ten per

L

cent. of alcohol, there was little or no difference; above that proportion, the retarding power was manifest, until, with fifty per cent. of alcohol,. digestion was seriously interrupted.

"These experiments, therefore, indicate that ardent spirits, as usually employed dietetically by temperate persons, act as pure stimulants to gastric digestion, causing an increased flow of gastric juice, and stimulating the muscular contractions of the viscus, and so accelerating the speed of the digestive processes."

Wines.—Regarding the action of wines he says :—" If we consider the copious proportions in which hock and claret are used dietetically, it becomes evident that their retarding effect on peptic digestion is often brought into play. . . . On the other hand, the more sparing use of these wines, a glass or two with dinner or luncheon, would evidently not produce any appreciable retardation of peptic action, but would, like corresponding doses of sherry, act as pure stimulants. In both these instances, as in some others, it seems to be indicated that by adjusting the quantities we may elicit diverse effects. With large quantities we may obtain retardation ; with small quantities we may obtain acceleration of gastric digestion " (*Roberts*).

Experience. as well as experimental evidence, show that the amount of alcohol taken by persons in good health in the twenty-four hours should not exceed the equivalent of one or one and a half ounces of absolute alcohol, and this, too, whether the beverage be wine, beer, or spirits.

Alcohol in Sickness.—During illness this amount will often have to be exceeded, but it is open to grave doubt whether the very large quantities sometimes employed are not useless, or even hurtful. The quality of the beverage selected, and the tolerance of the individual constitution, as well as the previous habits of the person with regard to the use of alcohol, have all to be taken into account in coming to a decision in particular cases.

Effects of Excess.—The evil effects that follow the continued use of alcoholic stimulants in excess are so well known as hardly to need mention. Digestion becomes seriously disordered; there are all the symptoms of chronic gastric catarrh, with usually morning vomiting. The heart is weak and the circulation feeble. Fatty degeneration and fibroid contraction of different organs are found in the later stages. The mental faculties are greatly impaired. The will is weak, and the impulses are uncertain. All this may take place without the individual ever taking enough at a time to make him intoxicated, or being, in the ordinary sense of the term, a drunkard. Occasional excess in alcohol does not produce such marked effects, as the system has time to recover itself between the bouts.

Alcohol with Food.—It has been already said that alcohol is not necessary to persons in perfect health, but to those who are of more feeble habit of body, with slow or imperfect digestive power, alcohol in limited amount is certainly often of use. It should, however, never be taken by such persons early in the day, or at any time as a stimulant to work upon, and it should be taken with or just after food, not upon an empty stomach. This is a good general rule; but it must be admitted that the local action of a very small quantity of alcohol upon the nerves of taste and upon the stomach may occasionally be applied with advantage at the beginning of a meal by those persons who, after some unusual fatigue, feel too tired to eat. A sip of wine or of dilute spirits and water at the commencement of the meal sometimes makes all the difference to the food being enjoyed and digested. The habitual use of such appetisers as sherry-and-bitters before dinner cannot, however, even on this ground, be recommended. Persons of unstable nervous system should avoid alcohol altogether.

The value of alcoholic beverages in acute disease has already been discussed (p. 141). Some other points may here be summarised.

Summary.—A small or moderate amount of alcohol is useful in all cases where the action of the heart is weak. This may occur either in the later years of life, or during convalescence from some acute illness, as well as during the course of acute disease, when there are signs of heart failure. The stimulant will, in these circumstances, help to sustain life, and so tide over a time of danger; and it is besides useful in so far as it increases appetite and aids the digestion of food.   Alcohol without food is decidedly dangerous, and life cannot be continued for any length of time upon it alone.

CHAPTER XIX.

PREPARED AND PREDIGESTED FOODS.

PREPARED AND PREDIGESTED FOODS.

SOME patients have difficulty in digesting starch, and in those cases where this defect exists the addition of malt to the farinaceous substance is very useful, partly on account of the nutritive qualities of the malt itself, but mainly because of the diastase and the action which that principle has in converting starch into dextrine and sugar.

Use of Malt.—Malt may be employed either in powder or in the form of an extract, and may be added to farinaceous foods, such as milk-pudding, just before they come to the table, and after they have cooled down to the temperature at which they can comfortably be eaten. If the malt extract be added while the temperature of the food is still high, the effect is lost, as the activity of the diastase is stopped by the heat. The malt may also, if preferred, be taken along with the food or immediately after it, but is comparatively useless if taken some considerable time after the food. Of prepared foods containing malt in combination with farinaceous substances, many varieties are now sold.

Prepared Foods.—Many preparations usually named "foods for invalids and infants," and of which the basis is baked flour of some sort, with malt, and with or without dried milk, can be obtained. While these foods hold a

165

useful place in the dietaries of the two classes for whom
they are intended, they are, it is to be feared, too often
used as substitutes for, instead of adjuncts to, milk.   Those
foods that contain milk have, of course, the higher nutritive value, and those containing sugar and malt are most
valuable as additions to milk.   Without milk and sugar,
they form by themselves a very insufficient food, but they
may be added to broths or beef-tea, and given as useful
alternatives to milk.

The predigestion of starch by heat was recognised before
the introduction of prepared farinaceous foods, as is seen in
the use of "tops and bottoms" for infants' food.   A further
great advance was made by the addition of malt to these
foods, and a typical one early introduced is "Liebig's food
for infants and invalids," containing malt powder, milk,
baked wheaten flour, and saline matter.   Another is that
prepared by Messrs. Allen, Hanburys, & Co., of which they
give the following analysis.

Analysis of Allen & Hanburys' Food.—One pound of
malted food contains :—

Starch rendered soluble by diastase, together with dextrine, maltose, and other soluble substances	13 oz.	130 grs.
Albuminoids	1 „	420 „
Water		143 „
Fat		60 ,
Cellulose		123 „
Undetermined substances		44 „
Mineral matter = phosphates, &c.		72 „

For one part of flesh-formers there are six parts of heat-
producers, reckoned as starch.

Kepler's Extract of Malt.—A reliable form is the Kepler
extract of malt.   A small teaspoonful added to a farinaceous pudding or a plate of oatmeal porridge so acts upon
the starch that in a comparatively few minutes liquefaction
takes place.   When symptoms of mal-nutrition are present,

and especially when there is reason to believe that the
system has difficulty in dealing with amylaceous substances,
the malt extract should be given a fair trial.

Predigestion of Foods.—Predigestion of food is necessary
in cases where the digestive powers are so enfeebled that
suitable food in ordinary forms cannot be digested. In
such circumstances predigested foods are undoubtedly useful.
In functional disorders they are of doubtful value, and they
should not be resorted to until ordinary means have been
tried and have failed.

In organic diseases of the stomach and bowels they are
very valuable, but they are useless, if not injurious, where
it is the secondary digestive processes that are at fault. So
much for the digestion of starch.

Pepsin and Pancreatic Ferments. — Next come pepsin
and the pancreatic ferments. Pepsin being the active principle of the gastric juice, retains its activity only in the
presence of an acid; hence the addition of a little dilute
hydrochloric acid frequently aids the action of pepsin.
Pepsin should be given at or just after those meals at
which animal food is taken, since it acts upon albuminoids,
and not upon farinaceous substances, nor upon fats. Pepsin
is useful in cases of weak digestion; for example, in the aged,
or in persons recovering from acute diseases. It may be
used in the form of powders or tabloids, which are very convenient, or in the liquid form. Of the last named, a useful
combination is the acid glycerine of pepsin.

Pepsin with dilute hydrochloric acid may be used to
predigest proteid foods, but it is found more convenient in
practice to employ liquor pancreaticus, and to Sir William
Roberts we are indebted for much valuable information of a
practical sort as to its use. The pancreatic fluid extract acts
in an alkaline medium only, and to meet this Sir William
Roberts, in his experiments, combined an alkali with liquor
pancreaticus, and gave the extracts some time after a meal,
gastric acidity being then considerably reduced.

Benger's Liquor Pancreaticus.—To predigest foods, an ordinary extract of the pancreas of the pig, made by maceration of that organ in dilute spirit, can be used, but it is troublesome to prepare, and Benger's liquor pancreaticus is a very convenient form. In the case of milk, it is necessary to dilute it first, so as to prevent firm curdling, and an alkali must be added to prevent coagulation in the final boiling.

The full instructions are the following:—

Peptonised Milk.—"A pint of milk is diluted with a quarter of a pint of water and heated to a temperature of about 140° Fahr. Should no thermometer be at hand, the diluted milk may be divided into two equal portions, one of which is heated to the boiling-point and added to the cold portion, when the mixture will be of the required temperature. Two teaspoonfuls of the liquor pancreaticus and ten grains of bicarbonate of soda are then added to the warm milk. The mixture is poured into a covered jug, and the jug is placed in a warm situation under a 'cosey,' in order to keep up the heat. At the end of an hour or an hour and a half the product is boiled for two or three minutes. It can then be used like ordinary milk."

Peptonised foods do not keep long without further change, and it is better, therefore, not to peptonise at one time more than is required for twelve hours' use. More or less liquor pancreaticus can be used, according to the extent to which it is wished to carry the peptonising process. If carried to the full extent, a disagreeable bitter taste is developed.

Peptonised Gruel.—Any ordinary farinaceous substance, such as flour, oatmeal, or arrowroot, may be used. "The gruel should be well boiled, and made thick and strong. It is then poured into a covered jug, and allowed to cool until it becomes lukewarm. Liquor pancreaticus is then added in the proportion of a dessertspoonful to the pint of gruel, and the jug is kept warm under a 'cosey' as before. At the end of a couple of hours the product is boiled and strained."

Peptonised Milk Gruel.—"A strong gruel of any farinaceous substance is made. To the boiling gruel add a like quantity of cold milk. The temperature of the united milk and gruel will be about 125° Fahr. Add about two teaspoonfuls of liquor pancreaticus and ten grains of bicarbonate of soda (a small quantity of light magnesia has been suggested as an alternative alkali). Place this in a covered jug under a 'cosey' for an hour and a half, then boil for two minutes, and strain."

Peptonised Beef-Tea.—"A pound of finely-minced lean beef is mixed with a pint of water, and ten grains of bicarbonate of soda are added. The mixture is then simmered for an hour and a half in a covered saucepan. The resulting beef-tea is decanted off into a covered jug. The undissolved beef residue is then beaten up with a spoon into a pulp, and added to the beef-tea in the covered jug. When the mixture has cooled down to about 140° Fahr. (or when it is cool enough to be tolerated in the mouth), a tablespoonful of the liquor pancreaticus is added, and the whole well stirred together. The covered jug is then kept warm under a 'cosey' for two hours, and agitated occasionally. At the end of this time the contents of the jug are boiled briskly for two or three minutes, and finally strained. The product is then ready for use."

Peptonised Soups, Jellies, and Blanc-Mange.—"Soups may be prepared in two ways. The first way is to add what cooks call 'stock' to an equal quantity of peptonised gruel or peptonised milk-gruel. A second and better way is to use peptonised gruel, which is quite thin and watery, instead of simple water, for the purpose of extracting the soluble matters of shins of beef and other materials employed in the preparation of soups. Jellies may be prepared by simply adding the due quantity of gelatine or isinglass to hot peptonised gruel, and flavouring the mixture according to taste. Blanc-mange may be made by treating peptonised milk in a similar way, and then adding cream. In preparing all these dishes, it is absolutely necessary to complete the operation of peptonising the gruel or the milk, even to the final boiling, before adding the stiffening ingredient; for if pancreatic extracts be allowed to act on the gelatine, the gelatine itself undergoes a process of digestion, and its power of setting or cooling is therefore utterly abolished" (*Sir William Roberts*).

Fairchild's Peptonising Powders (Zymine).—These are very useful for peptonising milk, and especially for use in the case of infants brought up by " bottle."

"Into a clean quart bottle pour a pint of fresh milk, one quarter of a pint of cold water, and one peptonising powder. Set this in water as hot as the hand can bear for thirty minutes, shaking occasionally. It may now be used."

When not for immediate use, it should be boiled for two or three minutes, or put on ice to prevent the milk from becoming bitter. Keep in a cool place.

Half an hour is as long as the process should be allowed to proceed, or the product will be unpleasantly bitter. Boiling is necessary to prevent peptonisation being carried out to the full extent, and thus leaving nothing for the digestive organs to do.

Beef-tea may also be peptonised by the use of these powders.

NUTRIENT ENEMATA.

Occasions for the Use of Nutrient Enemata.—However undesirable it may be to depend upon nutrient enemata alone for any long-continued support of the body, there can be no doubt that they are most valuable in certain circumstances, by enabling the patient to hold out and to maintain his strength until the crisis that called for their use has passed. Such crises arise sometimes in the course of acute gastric catarrh, in gastric ulcer, in ulceration of the small bowel, and in other circumstances when the stomach is intolerant of all food. Again, when very little

food can be taken and retained by the stomach, the balance
of nutrition may be kept up by the addition, twice a day,
of a nutrient enema.

That the large bowel possesses, to a certain extent, powers
of absorption is not now questioned, but it must be remembered that, under ordinary circumstances, the substances
upon which its absorptive capabilities are exercised have
been very much altered and prepared for absorption, by the
action of the gastric and intestinal juices in their passage
down the bowel. In other words, the foods have been, to a
great extent, digested before they reach the colon.

Use of Prepared and Predigested Foods.—This fact points
strongly to the use of prepared and predigested foods in
nutrient enemata. It may very well be doubted to what
extent the old-fashioned enema of beef-tea, with a beaten-
up egg, and some milk and brandy, was really absorbed.
So far as the use of prepared foods, however, is concerned,
we have every reason to believe that their predigestion
renders them, to a large extent, capable of absorption by the
large bowel. Sir William Roberts strongly recommends
the use of the liquor pancreaticus as an addition to nutrient
enemata, being convinced by his experience that it is very
valuable for that purpose. A dessertspoonful should be
added to an enema of beef-tea or of milk-gruel shortly
before it is injected into the bowel. The prepared farinaceous foods may likewise be used in this way, and also
beef peptones and malt extracts, with the addition, if necessary, of wine or spirit.

Some practical points in regard to the administration of
nutrient enemata must be noticed.

Rules.—(*a.*) A nutrient enema must always be small in
bulk. About two ounces is an average quantity for an
adult; but if a long flexible tube can be easily passed well
up the bowel, twice or three times that quantity may sometimes be comfortably retained.

(*b.*) A nutrient enema must be injected very gently and

slowly, else contractions of the bowel and the evacuation of its contents will certainly be set up.

(*c.*) The temperature of the enema is important. It must be lukewarm, not hot, and certainly not cold.

(*d.*) Great care must be exercised in the introduction of the syringe. It must be passed gently into the bowel, and not pressed upon the mucous membrane each time at the same spot. If this rule be neglected, there is great risk that abrasion and ulceration will be set up. This point is, of course, of special importance where the use of enemata has to be continued for a length of time.

(*e.*) Vaseline or oil should be used to lubricate the tube, and, at the same time, the anus, to prevent irritation. If the rectum and anus become irritable, the nutrient enemata cannot be retained, and this method of feeding may thus fail us at a critical moment.

(*f.*) A few drops of laudanum added to a nutrient enema aids its retention, and gentle pressure, by means of a soft pad over the anus, is also helpful in cases where an enema is retained with difficulty.

(*g.*) Care must be taken to see that the rectum does not become loaded with fæces. Occasional washing out of the bowel with tepid water is useful.

CHAPTER XX.

INFLUENZA.

THIS is not the place to discuss the pathology of the disease to which the name of influenza has become attached, nor have we space to enter upon a consideration of the etiology of the affection, but the importance of dietetic treatment in all forms of the disease admits of no question.

Some cases require a diet similar to that which we should advise for patients suffering from any specific fever, but in others special measures for the support of the strength are undoubtedly necessary. The tendency is towards asthenia, and this is more marked in some cases than in others, though weakness is always a more or less prominent symptom. The resulting muscular prostration and depression of the nervous system are in most cases out of all proportion to the apparent severity and duration of the attack. Many patients complain of inability to walk any distance with comfort, or without feeling exhausted, and this is usually most marked in those who have not been able, during the attack, to take nourishment well. In others the most notable persisting symptom is feeble action of the heart, with a tendency to fainting after exertion, or if they have gone long without food.

Three different varieties, at least, may be distinguished, according as the brunt of the attack has fallen upon the lungs, upon the gastro-intestinal mucous membrane, or upon the nervous system.

In the first group there is bronchial catarrh, with elevation of temperature and some prostration, but fortunately without, as a rule, great repugnance to food. In these cases

169*

we should advise a diet like the following at the commencement :—

Six to eight ounces of milk, with one or two ounces of barley water, or of soda water, may alternate at intervals of two-and-a-half to three hours with like quantities of mutton-broth or chicken-tea or beef-tea. The broth or beef-tea should be thickened with some farinaceous substance (see p. 140). After a few days, when in uncomplicated cases temperature will have fallen, a diet of four small meals a day, with intermediate supplies of light nourishment will be suitable, thus :—

Early in the morning, a cup of milk flavoured with tea.

8.30. A lightly boiled egg, or a piece of fresh white fish, with toast and butter and one cup of weak tea with milk, or coffee and milk or cocoa.

11.30. A cupful of broth or beef-tea, with fingers of toast, or thickened with some biscuit (crumbled).

1.30. Fresh white fish or chicken, or other bird, with a little mashed potato and well-boiled green vegetable : some simple pudding.

4.30. A cup of weak tea with plenty of milk, toast and butter.

7.30. A meal like the midday one.

10.30–11. A cupful of prepared farinaceous food, or of gruel or arrowroot.

Stimulants.—The amount will vary in different cases, according to the state of the pulse and the general strength of the patient. Where there is no special indication of weakness, a tablespoonful of spirit in water, with the midday and with the evening meal, will suffice. Others will require and will benefit by a larger allowance. A dessertspoonful at 11.30 and another at 7.30 in water, or in effervescing water, and one in the food at bedtime, will not in these cases be too much. As convalescence progresses a couple of glasses of champagne at the midday and evening meals may take the place of the spirits.

In the group of cases, where gastro-intestinal catarrh is a prominent symptom, careful dietary is very essential. Food must be bland, non-irritating, and easily digested, leaving little residue, and not such as is likely to set up fermentative changes. For details the reader is referred to pp. 39 and 40.

In the third group of cases, where the nervous system is profoundly affected, patients can take and can bear large quantities of food. In some cases, indeed, of this type, feeding must be kept up regularly through the night, sleep being obtained only after food has been taken.

A characteristic of many cases in this group is the absence of any rise of temperature, or other sign of febrile disturbance. The temperature, indeed, often falls below normal, and this is most noticeable when any considerable time has elapsed since the last supply of food has been taken.

In the class of cases we are now considering, it is not necessary to restrict the patient entirely to liquid nourishment. Solids may be given almost, if not quite, from the commencement, though of course food must be very simple. There should be three good meals a day, with some intermediate supplies between them, and light food should be given several times during the night.

Breakfast.—A couple of eggs, or egg and bacon, or white fish, with bread, or toast and butter; a cup of café-au-lait or of cocoa.

In the forenoon a small cup of soup or of good beef-tea thickened with some farinaceous material.

Luncheon.—Any light tender meat—chicken, game, or mutton—with a little mashed potato; any simple pudding.

Beverage.—A couple of glasses of good claret or of champagne, or from one to two tablespoonfuls of whisky in a little effervescing water.

Afternoon Tea.—Tea, with plenty of milk, or cocoa; bread and butter or toast and butter.

Dinner.—A meal like luncheon.

Beverage.—A pint of claret or of champagne. Before settling for the night a cupful of prepared farinaceous food or of gruel, with a tablespoonful of brandy or whisky in it.

During the night milk or soda water, with a dessert-spoonful of brandy in it, or beef jelly or beef-tea, warmed by being kept under a cosey, should be given once or twice.

Stimulants.—In the above directions the amount of stimulant mentioned is not more than would be required in an average case of the kind, but in many a considerably larger quantity may be given, not only without harm, but with great benefit.

APPENDIX.

SICK-ROOM COOKERY.

GENERAL CONTENTS—COOKING FOR INVALIDS: Beef Essences—Beef-Tea—
Nutritious Beef-Tea—Beef-Tea with Oatmeal—Mutton Broths—Chicken-
Tea—Calf's-Foot Broth—Mutton Broth—Veal Broth—Egg and Brandy
—Egg and Sherry—Caudle—Another Caudle—Milk and Isinglass—
Arrowroot—A Gruel—Oatmeal Gruel—Tamarind Water—Arrowroot
and Black-Currant Drink—Cream of Tartar (*Potus Imperialis*)—Rice
Water—Snow Pudding—Lemonade—Milk Lemonade—Rice and Milk
—Oatmeal Porridge—Milk Porridge—Whole Meal Porridge—Irish Moss
—Toast Water—Barley Water—Milk, Eggs, and Brandy—Port-Wine
Jelly—Bread Jelly—Wine Jelly—Chicken Panadas—Game Panada—
Nourishing Soup—Tapioca Soup with Cream—Purée of Potatoes—Cream
of Barley—Maccaroni with Milk—Maccaroni (Stewed in Stock)—Lamb's
Head—Cow-Heel Fried—Ox Palates—Sweetbread—Sweetbread with
White Sauce—Calf's Head—Tripe—Breast of Lamb with Vegetables—
Kedgeree—Fish Soup—Calf's-Foot Jelly—Blanc-Mange — Arrowroot
Pudding—Custards—Rice Pudding—Rice Cream—Corn-Flour—Hominy
Pudding—Blanc-Mange—Cream — A Ripe Fruit Cream—Chocolate
Cream—Summer Fruit Pudding—Apple Charlotte—Charlotte Russe—
Omelet Soufflé—Omelet Savoury—Peptonised Milk—Peptonised Gruel
—Peptonised Milk Gruel—Peptonised Soups, Jellies, and Blanc-Manges
—Peptonised Beef-Tea—Peptonised Enemata—Whey—White Wine
Whey—"Tops and Bottoms in Milk."

1. Beef Essence.

Take one pound of gravy beef, free from fat and skin. Chop
it up very fine; add a little salt, and put it into an earthen
jar with a lid; fasten up the edges with a thick paste, such
as is used for roasting venison in, and place the jar in the
oven for three or four hours. Strain through a coarse sieve,
and give the patient two or three teaspoonfuls at a time.

M

2. Beef Essence.

Cut up in small pieces one pound of lean beef from the sirloin or rump, and place it in a covered saucepan, with half a pint of cold water, by the side of the fire for four or five hours then allow it to simmer gently for two hours. Skim it well, and serve.

3. Beef-Tea.

Cut up a pound of lean beef into pieces the size of dice; put it into a covered jar with two pints of cold water and a pinch or two of salt. Let it warm gradually and simmer for a couple of hours, care being taken that it does not reach the boiling-point.

4. Nutritious Beef-Tea.

To a pint of beef-tea or mutton broth (not too strong) add two tablespoonfuls of powdered biscuits or bread-crumb; boil for five minutes, stirring well all the time.

5. Beef-Tea with Oatmeal.

Mix two tablespoonfuls of oatmeal, very smooth, with two spoonfuls of cold water; then add a pint of strong boiling beef-tea. Boil together for five or six minutes, stirring it well all the time. Strain it through a sieve, and serve.

6. Mutton Broth.

Cut one pound of lean mutton into dice; to this put one quart of cold water, then let it simmer on the hob for three hours; take off the scum as required, and add a pinch of salt. Strain off the fluid, let it stand till it is cold, then remove the fat, if any

7. Mutton Broth.

Two or three pounds of neck of mutton, two pints of water, pepper and salt, half a pound of potatoes, or some pearl barley.

Put the mutton into a stewpan, pour the water over it, add pepper and salt. When it boils, skim carefully; cover, the pan and let it simmer gently for an hour. Strain it, let it get cold, and then remove all the fat. When required for use, add some pearl barley or potatoes in the following manner :—Boil the potatoes, mash them very smoothly, so that no lumps remain. Put the potatoes into a pan, and gradually add the mutton broth, stirring it till it is well mixed and smooth ; let it simmer for five minutes, and serve with fried bread.

8. Chicken-Tea.

Cut up a fowl in small pieces. Put it into an earthen vessel with some salt and three pints of water ; let it boil three hours, set it to cool, then take off the fat.

9. Calf's-Foot Broth.

One calf's foot, three pints of water, one small lump of sugar, the yolk of one egg.

Stew the foot in water, *very gently*, till the liquor is reduced to half; remove the scum, set it in a basin till quite cold, then take off every particle of fat. Warm up about half a pint, adding the sugar; take it off the fire for a minute or two, then add the beaten yolk of the egg. Keep stirring it over the fire till the mixture thickens, *but do not let it boil*, or it will be spoiled.

10. Mutton Broth.

Take a pound of mutton, cut it up into small pieces, take a thick slice of bread, toast one-half of it a rich brown colour, crumble the other half, and put it into two quarts of cold water with any flavouring preferred, or even without any ; boil gently to a pint and a half, and strain. Take off the fat when cold, if it be objected to.

11. Veal Broth.

A knuckle of veal, two cow-heels, twelve pepper-corns, a glass of sherry, and two quarts of water.

Stew all the ingredients in an earthen jar six hours. Do not open it till cold. When wanted for use, skim off the fat and strain it. Heat as much as you require for use (*Ringer*).

12. Egg and Brandy.

Beat up three eggs to a froth in four ounces of cold water, add two or three lumps of sugar, and pour in four ounces of brandy, stirring all the time. A portion of this to be given at a time.

13. Egg and Sherry.

Beat up with a fork an egg till it froths, add a lump of sugar and two tablespoonfuls of water; mix well, pour in a wineglassful of sherry, and serve before it gets flat.

14. Caudle.

Beat up an egg to a froth, add a wineglassful of sherry and half a pint of gruel; flavour with lemon-peel and nutmeg, and sweeten to taste.

Another Caudle.

Mix well together one pint of cold gruel with a wineglassful of good cream; add a wineglassful of sherry and a tablespoonful of noyeau, and sweeten with sugar-candy.

15. Milk and Isinglass.

Dissolve a little isinglass in water, mix it well with half a pint of milk, then boil the milk, and serve with or without sugar, as preferred.

16. Arrowroot.

Mix two teaspoonfuls of the best arrowroot with half a wineglassful of cold water; add a pint of boiling water; put it into an enamelled saucepan, and stir over the fire for three minutes. Sweeten with three teaspoonfuls of sifted loaf-sugar.

17. A Gruel.

Beat up an egg to a froth; add a wineglass of sherry; flavour with a lump of sugar, a strip of lemon-peel, and a little grated nutmeg. Have ready some gruel, very smooth and hot; stir in the wine and egg, and serve with sippets of crisp toast. Arrowroot may be made in the same way.

18. Oatmeal Gruel

Put a pint of boiling water into a saucepan; into this stir a couple of tablespoonfuls of oatmeal until quite smooth; let this boil well for ten or fifteen minutes; season with salt, then strain through a strainer.

19. Tamarind Water.

Boil two ounces of tamarinds with a quarter of a pound of stoned raisins in three pints of water for an hour; strain, and when cold it is fit for use.

20. Arrowroot and Black-Currant Drink.

Take two large spoonfuls of black-currant preserve; boil it in a quart of water; cover it, and stew gently for half an hour, then strain it, and set the liquor again on the fire; then mix a teaspoonful of arrowroot in cold water, and pour the boiling liquor upon it, stirring meanwhile; then let it cool.

21. Cream of Tartar (*Potus Imperialis*).

Put half an ounce of cream of tartar, the juice of one lemon, and two tablespoonfuls of sifted sugar into a jug, and pour over a quart of boiling water. Cover till cold.

22. Rice-Water.

Wash three ounces of rice in several waters, and then put it into a clean stewpan with a quart of water and one ounce of raisins; boil gently for half an hour, and strain into a jug. When cold it is ready for use.

23. Snow-Pudding.

Put into half a pint of cold water half a package of gelatine. Let it stand one hour; then add one pint of boiling water, half a pound of sugar, and the juice of two lemons. Stir and strain, and let it stand all night. Beat very stiff the whites of two eggs and beat well into the mixture. Pour into a mould.

24. Lemonade.

Express the juice of a large-sized lemon and strain it. Put it into a jug, adding about a third of the rind cut thin, and four or five pieces of lump sugar. Pour over this one pint of boiling water, and cover it over for two hours. Strain, and when cool it is ready for use.

25. Milk Lemonade.

To a pint of boiling water add six ounces of loaf sugar, six ounces of lemon-juice, a like quantity of sherry, and sixteen ounces of cold milk. Stir thoroughly, and strain through a jelly-bag.

26. Rice and Milk.

To a quart of milk add a quarter of a pound of rice, which has been well washed; simmer for an hour, stirring very frequently. Flavour with cinnamon or lemon-peel, and just before serving sweeten to taste.

27. Oatmeal Porridge.

Boil some water quickly in a saucepan, and add to it a pinch of salt. Into the water sprinkle some oatmeal, stirring briskly all the time. Let it simmer for at least half an hour. The porridge may be made thick or thin, according to taste.

28. Milk Porridge.

Into half a pint of water put two ounces of oatmeal. Let it soak for ten or twelve hours; then pass it through a sieve into a saucepan; boil for half an hour, and before serving add five or six ounces of milk.

29. Whole-Meal Porridge.

Into a quart of boiling water sprinkle slowly half a pound of wheat-meal, and boil till thoroughly soft. It may be served like ordinary porridge with milk or sugar, or marmalade or treacle, or Kepler's extract of malt.

30. Irish Moss.

Wash well an ounce of Irish moss, and then let it soak for four or five hours in a quarter of a pint of cold water. Add half a pint of milk and half a pint of water (or a pint of milk); boil from five to seven minutes, and strain through muslin. It may be flavoured with cinnamon or vanilla, or simply sweetened with sugar, and when cool will set firmly.

31. Toast-Water.

Boil a quart of water, and pour it upon a good-sized piece of crumb of bread, which has been well toasted before a clear fire, until it becomes very crisp and of a dark-brown colour (not black). Let it steep for half an hour, and then decant it into a jug.

32. Barley-Water.

Put an ounce of barley-water into an enamelled saucepan with a quart of cold water, and boil for two hours. Stir it, and skim it occasionally. Strain through muslin, and sweeten with a little sugar-candy. The juice of a lemon, strained, may be added.

33. Milk, Egg, and Brandy.

Scald some new milk, but do not let it boil. It ought to be put into a jug, and the jug should stand in boiling water. When the surface looks filmy, it is sufficiently done, and should be put away in a cool place in the same vessel. When quite cold, beat up a fresh egg with a fork in a tumbler, with a lump of sugar; beat quite to a froth, add a dessert-spoonful of brandy, and fill up the tumbler with scalded milk.

34. Port-Wine Jelly.

Put into a jar one pint of port-wine, two ounces of gum-arabic, two ounces of isinglass, two ounces of powdered white sugar-candy, a quarter of a nutmeg grated fine, and a small piece of cinnamon. Let this stand closely covered all night. The next day put the jar into boiling water, and let it simmer till all is dissolved; then strain it, let it stand till cold, and then cut it up into small pieces for use (*Ringer*).

35. Bread-Jelly.

Take three or four slices of bread, remove the crust, and toast them to a light brown colour. Put the bread into a pan with two pints of cold water, and simmer for an hour and a half or two hours; strain through muslin, sweeten, and flavour with a little wine and lemon-juice. Pour into a mould, and turn out when required for use.

36. Wine-Jelly.

Into an enamelled saucepan put a pint of cold water; add an ounce and a quarter of Russian isinglass and three ounces of sugar. Let this boil gently till the isinglass is dissolved, removing the scum from time to time. Strain, and add the juice of one lemon or of two oranges, and a third of a pint of sherry. Pour into a mould, and turn out when required.

37. Chicken Panada (1).

Take the flesh from the breast of a fresh roasted chicken; soak the crumb of a French roll or a few rusks in hot milk, and put this in a clean stewpan, with the meat from the chicken previously reduced to a smooth pulp by chopping it and pounding it in a mortar; add a little chicken-broth or plain water, and stir the panada over the fire for ten minutes.

38. Chicken Panada (2).

Half a chicken is required for one panada. Take the flesh off the bones, cut it into small pieces, and put them into a gallipot, sprinkling over them half a salt-spoonful of salt. Cover the top of the gallipot with paper, tied down, and stand it in a saucepan half-full of boiling water. Let it simmer for two hours. Take out the pieces of chicken with a spoon, put them in a mortar, and pound them to a pulp. Pass the pounded chicken through a sieve, adding a little of the liquor to make it pass more easily. Stir into the basin with the pulp one tablespoonful of cream.

39. Game Panada.

Prepare in the same manner as the foregoing. The breast of a pheasant or partridge may be used in place of chicken.

40. Nourishing Soup.

Take a pound of good lean beef and a pound of mutton, and cut them into pieces the size of dice. Take also a calf's foot and split it. Put them into a jar with two quarts of cold water, and let them simmer in the oven for five or six hours, adding about the middle of that time another quart of water, and some simple seasoning. When the quantity is reduced to one and a half quarts, take it out of the oven and strain. When cold, remove the fat. This soup may be taken cold, or may be warmed up with a little pepper and salt.

41. Tapioca Soup with Cream.

Take a pint of white stock and pour into a stewpan. When it comes to the boil, stir in gradually one ounce of prepared tapioca. Let it simmer slowly by the side of the fire until the tapioca is quite clear. Put the yolks of two eggs into a basin, with two tablespoonfuls of cream. Stir with a wooden spoon, and pour through a strainer into another basin. When the stock has cooled, add it by degrees to the mixture, stirring well all the time, so that the eggs may not curdle. Pour it back into the stewpan, and warm before serving. Add pepper and salt to taste.

41a. Purée of Potatoes.

Take one pound of potatoes, peel them, and cut them in thin slices, also two leaves of celery well washed. Put an ounce of butter in a stewpan, add the potatoes and celery. Put the stew-pan on the fire, and let the vegetables sweat for five minutes, taking care they do not discolour. Pour into the stewpan one pint of good white stock; stir frequently to prevent burning. Let it boil gently till the vegetables are quite cooked. Heat half a pint of the white stock in another stewpan. Pass the contents of the first stewpan through a sieve into a basin, adding by

degrees the half-pint of hot white stock, which will enable it to
pass through more easily. Pour the purée back into the stewpan, and season according to taste, adding a quarter of a pint of
cream. Stir smoothly till it boils.

42. Cream of Barley.

Take half a pound of veal cutlet; pare off all the fat, and cut
the lean into small pieces. Put this into a saucepan with one
pint of cold water; add half an ounce of well-washed barley that
has soaked for an hour or two in cold water, and half a salt-
spoonful of salt. Let it boil for two hours. Strain off the
liquor, and put the meat and barley in a mortar. Pound them
together, and then rub them through a sieve, pouring on the
sieve a little of the liquor. Stir in two tablespoonfuls of cream.

43. Maccaroni with Milk.

Take half a pound of maccaroni; wash it, and put it in a saucepan of cold water with a tablespoonful of salt. Let it boil for
half an hour; then pour out the water, and add to the maccaroni
one quart of milk, and let it simmer gently for an hour.

44. Maccaroni (*Stewed in Stock*).

Take half a pound of maccaroni; put it into a saucepan of cold
water with a dessert-spoonful of salt. Let it boil for ten minutes,
and then put the maccaroni into a cullender and run some cold
water on it. After draining it, turn it on a board and cut it
into small pieces. Put the maccaroni thus cut up into a saucepan with one pint of good white stock. Bring it to the boil, and
let it simmer gently for ten minutes. Season with pepper and
salt to taste, and serve on a hot dish.

45. Lamb's Head.

Take out the brains and cut them up fine, and stew them in
a white sauce. Stew the head till quite tender. Take the meat
off the bones; lay it on the centre of a dish; egg and bread-
crumb it, and brown in an oven. Serve with brain-sauce poured
round the dish.

46. Cow-Heel Fried.

Stew the heel about eight hours until the meat leaves the bone easily; press it between two plates till cold and set; cut it into strips, roll them up; egg and bread-crumb them, and fry them. Or,—

Simply stewed, and the meat, when taken from the bone, served with good stock sauce.

47. Ox Palates.

Boil the ox palates about eight hours; then skin them, and serve with a good brown gravy.

48. Sweetbread.

Soak one sweetbread in water for an hour; then change the water and soak for another hour; then put them into boiling water and let them simmer for ten minutes, or rather more, till they become firm; drain, and when cool brush well over with beaten-up egg; roll them in bread-crumbs, dip again in egg, and add some more bread-crumbs; sprinkle over them a little oiled butter, and bake in an oven, basting them till they are sufficiently done, and of a bright brown colour. Serve on toast, with a little brown gravy round the dish.

49. Sweetbread with White Sauce.

Soak one sweetbread for an hour. After boiling for ten minutes, put it into cold water for twenty minutes. Pour over it, while in a saucepan, half a pint of good white stock; season according to taste. Simmer gently for half an hour; take it out and keep it warm in the oven. Let the sauce boil, thickened with a little flour and butter, and when quite smooth stir in two tablespoonfuls of cream. Place the sweetbread on a dish; add a teaspoonful of lemon-juice to the sauce, and pour it over.

50. Calf's Head.

Take half a calf's head; soak it well in-cold water. Parboil it in water, to which has been added a little salt, for nearly half an hour. Put it into a stewpan with seasoning; add a little

water, and let the whole boil gently for an hour and a half.
Serve with parsley-sauce poured over it, and small pieces of fat
bacon fried.   The tongue and brains may be served on the same
dish or separately.

51. Tripe.

Take half a pound of clean, fresh tripe, and wash it in cold
water; then cut into small squares, taking care to leave but little
fat. Put half a pint of milk into a stewpan, with a pinch of salt,
and one of castor-sugar; add a little mustard. Put the tripe
into the stewpan, let it boil up, skimming carefully. It should
now be allowed to simmer for three or three and a half hours,
and should be stirred frequently during that time. Mix a teaspoonful of cornflour with a little cold milk to a smooth paste,
and stir it in. After simmering for five minutes more, it may
be served on a hot dish, with the sauce poured over it.

52. Breast of Lamb with Vegetables.

Take the breast of tender lamb, cut away the skin and most of
the fat.   Divide it into small pieces, and sprinkle with flour.
Brown them in a stewpan with a little butter; add as much
warm water as will cover the meat. Flavour with vegetables to
taste, and add pepper and salt.   Remove the fat, and simmer for
an hour and a half.   Before simmering, a pint of young peas or
tender broad beans may be added.   If these are not in season,
pearl-barley answers well.

53. Kedgeree.

Take some cold cooked fish, break it well down. Boil for half
an hour a cupful of rice in white stock. Mix the fish with the
rice, and when the whole is heated stir in an egg.

54. Fish Soup.

Take the liquor in which fish has been boiled. Take also the
bones, skin, and other parts, and put them into two pints of the
liquor, and let it simmer till reduced to two-thirds of its bulk.
Strain, and season with vegetables, as well as salt and pepper and
a few drops of anchovy. When about to be served, stir in a
small cupful of boiling milk.

55. Calf's-Foot Jelly.

Take two calf's feet, split them, boil them in one quart of water; skim them thoroughly, and let them simmer for about four hours. Strain, and when cold remove the fat. Put into a stewpan with one pound of sugar the juice of six lemons, the rind of three, half a stick of cinnamon (bruised). Let this dissolve over a slow fire, and then add the whites of three eggs, well whipped up with five ounces of water, and continue until it begins to boil. Add half a pint of sherry; allow it to simmer gently for another quarter of an hour. Strain through a jelly-bag. Repeat the straining several times, and pour into moulds.

56. Blanc-Mange.

Put a quart of milk into a saucepan with three ounces of loaf-sugar, and one inch of cinnamon-stick, or a small piece of lemon-peel for flavouring. Let it boil. Put four tablespoonfuls of cornflour into a basin, and mix it smoothly with a little cold milk. When the milk in the saucepan is quite boiling, stir in the cornflour quickly, and let it boil for two minutes, stirring continually. Remove the flavouring, and pour into a mould. A little cream added to the milk is an improvement.

57. Arrowroot Pudding.

Take a dessert-spoonful of arrowroot, put it into a basin with the yolks of two eggs, a dessert-spoonful of milk and a teaspoonful of castor-sugar. Whip it smoothly into a paste. Take a saucepan, and put into it half a pint of cold milk. Watch the milk carefully till it boils. When quite boiling, pour it on to the arrowroot mixture, stirring all the time to get it quite smooth. Whip the whites of the two eggs, and add them to the mixture, stirring them lightly together. Pour the mixture into a buttered dish, and put it in a quick oven for ten minutes.

58. Light Pudding.

Make into a smooth paste a teaspoonful of flour, and pour over it six ounces of boiling milk, slightly flavoured, if desired.

Stir well, while adding a teaspoonful of sugar and a pinch of salt; when cold strain, and add it to a beaten-up egg. Pour the whole into a basin (buttered), and bake for twenty minutes in the oven.

59. Custard (1).

Boil a pint of new milk in an enamelled saucepan for two or three minutes, with three ounces of loaf-sugar, and the rind of half a lemon cut into thin slices. Allow it to cool for a few minutes. Beat up five eggs, and while stirring briskly add the milk to the eggs. Put the mixture again into the saucepan, and warm it over a gentle fire, constantly stirring till it begins to thicken. Strain through a fine sieve, and add two wine-glasses of good cream. The custard may be flavoured with brandy or with a liqueur.

60. Custard (2).

Take two eggs to a gill of milk, beat the eggs, and warm the milk before adding it to the eggs. Put it into a jug, and the jug into a pan of boiling water, and stir it till it thickens. It takes about a quarter of an hour.

61. Rice-Pudding.

Into half a pint of milk put two teaspoonfuls of rice well washed. Boil for two hours, stirring frequently, and sweeten with pounded sugar; take it off the fire and allow it to cool for a quarter of an hour. Then stir in the yolks of two eggs well beaten up. Bring it again to the boiling point, and boil for about one minute, stirring the while.

62. Rice-Cream.

Boil three ounces of rice in a quart of milk till quite soft; when cold, whip into it a teacupful of good cream, and half an ounce of isinglass melted in milk; when well whipped, pour into a mould with a hollow centre, and turn out when quite firm. Fill the centre with fresh stewed fruit or with any good preserve; whip up a pint of thick cream and pour it over the shape.

63. Cornflour.

Mix a dessert-spoonful of cornflour with four or five table-spoonfuls of milk. Boil half a pint of milk, and add this to the cornflour paste, stirring over the fire for four or five minutes; sweeten to taste, and, if allowable, add a tablespoonful of cream.

64. Hominy Pudding.

Soak for ten or twelve hours half a pint of hominy in a pint and a quarter of boiling water (the vessel should have a well-fitting cover). When thoroughly soaked, put the hominy into a pudding dish with a pint of milk, and bake in the oven for twenty or thirty minutes.

65. Blanc-Mange.

Take one ounce of pure isinglass, and dissolve it in a saucepan over a slow fire in one pint of milk. To this add half the rind of a lemon and half a pint of cream. After boiling for ten or fifteen minutes, take out the lemon rind, sweeten and flavour to taste. Before pouring into a mould, a little wine or brandy may be stirred in.

66. Cream.

Whip a pint of good cream to a froth. When thoroughly whipped, add a liqueur glass of brandy, or of a liqueur, six ounces of castor-sugar, and two ounces of isinglass. Mix thoroughly, and pour into a mould. When cold it is ready for use, and may be turned out.

67. A Ripe Fruit Cream (*e.g., Strawberry*).

Take a basket of strawberries, bruise the fruit, with six ounces of castor-sugar. Then rub through a hair sieve, add a pint of whipped cream and two ounces of pure isinglass; when thoroughly mixed, pour into a mould. Raspberries, &c., may be used in the same way.

68. Chocolate Cream.

Take three ounces of grated chocolate, and put it into a stew-pan, with half a pint of boiled milk, three ounces of sugar, and the yolks of four eggs. Stir on the fire till it thickens: pass through a hair sieve, add one ounce of gelatine, and when thoroughly mixed pour into a mould.

69. Summer Fruit Pudding.

Line a pudding-basin with bread and butter; have stewed fruit ready (red-currants and raspberries or strawberries are best because of their juice). Mix plain bread-crumbs with the fruit, and put into the basin. Turn it out and cover it with custard. The pudding must be made the day before it is used to fix it thoroughly, and to allow the bread to become saturated with the fruit.

70. Apple Charlotte.

Take two pounds of good cooking apples peeled, and cut into slices without the cores. Put the slices into a stewpan with sufficient sugar to sweeten them and quarter of a pint of water. Take the rind of a lemon and tie it together with a piece of string, and put it into the stewpan with the apples. Stir all on the fire till it boils, and the apples are reduced to about half their quantity. It will take from one to one and a half hours. Take out the lemon-peel; cut a slice of the crumb of bread; line the mould at the bottom and sides, having first dipped the slices into clarified butter. Pour the apples into the mould, and cover over with a round of bread like that lining the bottom of the mould. Bake in an oven for three-quarters of an hour.

71. Charlotte Russe.

Take a pint tin; line it inside with finger-biscuits, fitting the biscuits close to each other, and cutting off any points that remain above the tin. Put half an ounce of the best gelatine with a quarter of a pint of cold milk, and let it soak well. Take half a pint of double cream and whip it in a basin to a stiff froth; add a dessert-spoonful of sifted sugar, and flavour with vanilla, lemon, or orange. Put the stewpan on the fire and stir the gelatine till it is melted. Pour the melted gelatine through a sieve, and stir into the cream.

72. Omelet Soufflé.

Break two eggs; put the whites in one basin, the yolks in another. Put one teaspoonful of orange-flower water and one tablespoonful of castor-sugar into a stewpan. Let them boil quickly for three minutes. Then pour it into a cup to cool, adding to it the yolks of the eggs, and beat them to a cream; add a pinch of salt to the whites of egg and whip them to a stiff froth; add the whites to the mixture in the basin, and mix them together very lightly. Put half an ounce of butter into a frying-pan, and when the butter is hot pour in the mixture. Let it remain on a slow fire for about two and a half minutes; then put the pan into a hot oven for three or four minutes. Melt a dessert-spoonful of jam. Take the soufflée out of the oven, turn it on to a hot dish; spread the jam upon it and fold it over like a sandwich. Sprinkle it over with white sugar.

73. Omelet (*savoury*).

Made as above, omitting the jam, sugar, and orange-flower water, and adding salt and pepper, with parsley and other herbs.

74. Peptonised Milk.

" A pint of milk is diluted with a quarter of a pint of water, and heated to a temperature of about 140° Fahr. Should no thermometer be at hand, the diluted milk may be divided into two equal portions, one of which is heated to the boiling-point, and added to the cold portion, when the mixture will be of the required temperature. Two teaspoonfuls of the liquor pancreaticus and ten grains of bicarbonate of soda are then added to the warm milk. The mixture is poured into a covered jug, and the jug is placed in a warm situation under a ' cosey ' in order to keep up the heat. At the end of an hour or an hour and a half the product is boiled for two or three minutes. It can then be used like ordinary milk."

N

"The object of diluting the milk is to prevent the curdling which would otherwise occur, and greatly delay the peptonising process. The addition of bicarbonate of soda prevents coagulation during the final boiling, and also hastens the process. The purpose of the final boiling is to put a stop to the ferment action—when this has reached the desired degree—and thereby to prevent certain ulterior changes which would render the product less palatable. The degree to which the peptonising change has advanced is best judged of by the development of a peculiar bitter flavour, which is always associated with the artificial digestion of milk. The point aimed at is to carry the change so far that the bitter flavour is just perceived, but not unpleasantly pronounced."

75. Peptonised Gruel.

"Gruel may be prepared from any of the numerous farinaceous articles in common use,—wheaten flour, oatmeal, arrowroot, sago, pearl-barley, pea or lentil flour. The gruel should be well boiled, and made thick and strong. It is then poured into a covered jug, and allowed to cool until it becomes lukewarm. Liquor pancreaticus is then added, in the proportion of a dessert-spoonful to the pint of gruel, and the jug is kept warm under a 'cosey' as before. At the end of a couple of hours the product is boiled and strained."

76. Peptonised Milk-Gruel.

"First, a thick gruel is made from any of the farinaceous articles above mentioned. The gruel, while still boiling hot, is added to an equal quantity of cold milk. The mixture will have a temperature of about 125° Fahr. To each pint of this mixture two or three teaspoonfuls of liquor pancreaticus and ten grains of bicarbonate of soda are added. It is kept warm in a covered jug under a 'cosey' for an hour or an hour and a half, and then boiled for two or three minutes and strained. If the product has too much bitter flavour, a smaller quantity of the liquor pancreaticus must be used in the next operation. Invalids take this compound as a rule, if not with relish, at least without any objection.

"As it is impossible to obtain pancreatic extract of absolutely constant strength, the directions as to the quantity to be added must be understood with a certain latitude."

77. Peptonised Soups, Jellies, and Blanc-Manges.

" Soups may be prepared in two ways. The first way is to add what cooks call stock, and to an equal quantity of peptonised gruel or peptonised milk-gruel. A second and better way is to use peptonised gruel, which is quite thin and watery, instead of simple water, for the purpose of extracting the soluble matters of skins of beef and other materials employed in the preparation of soups.

" Jellies may be prepared by simply adding the due quantity of gelatine or isinglass to hot peptonised gruel, and flavouring the mixture according to taste.

" Blanc-manges may be made by treating peptonised milk in a similar way, and then adding cream. In preparing all these dishes, it is absolutely necessary to complete the operation of peptonising the gruel or the milk, even to the final boiling, before adding the stiffening ingredient. For if pancreatic extract be allowed to act on the gelatine, the gelatine itself undergoes a process of digestion, and its power of setting on cooling is therefore utterly abolished."

78. Peptonised Beef-Tea.

" A pound of thinly-minced lean beef is mixed with a pint of water, and ten grains of bicarbonate of soda are added thereto. The mixture is then simmered for an hour and a half in a covered saucepan. The resulting beef-tea is decanted off into a covered jug. The undissolved beef residue is then beaten up with a spoon into a pulp, and added to the beef-tea in the covered jug. When the mixture has cooled down to about 140° Fahr. (or when it is cool enough to be tolerated in the mouth), a table-spoonful of the liquor pancreaticus is added, and the whole well stirred together. The covered jug is then kept under a 'cosey' for two hours, and agitated occasionally. At the end of this time the contents of the jug are boiled briskly for two or three minutes and finally strained. The product is then ready for use."

79. Peptonised Enemata.

" Pancreatic extract is peculiarly adapted for administration with nutritive enemata. The enema may be prepared in the usual way with a mixture of milk and gruel, or milk, gruel, and beef-tea.   A dessert-spoonful of liquor pancreaticus is added to it just before administration " (*Sir William Roberts*).

80. Whey.

Into warm milk put a sufficient quantity of rennet to cause curdling, and strain off the liquid, which is then ready for use.

81. White-Wine Whey.

Boil a pint of milk, and while boiling add two wineglassfuls of sherry.   Strain through a fine sieve, and sweeten with sifted sugar.

82. "Tops and Bottoms" in Milk.

Over the bread in a basin pour a sufficiency of boiling water to soak it.   Let it stand for a minute or two, then strain off the water, sprinkle with a little sugar, and add hot milk.

83. Savoury Jelly.

Take half a chicken, one pound of neck of veal, one pound lean beef (from under the shoulder is the best part for beef-tea, &c.), cut the meat off the chicken, separate the joints, then cut all the meat, veal, beef, and chicken up very small; put the whole into an earthen mug, with two quarts of water, salt, pepper, a few herbs (whichever may be allowed); let them stew gently five or six hours in the oven; take off all fat with blotting-paper; strain through a hair-sieve into a mould.   When hot, give the patient a few tea-spoonfuls at a time.   This is a very nutritious jelly.

84. Barley-Gruel.

Take one ounce of pearl-barley, one ounce of rice, one ounce of sago.   Wash them all, then put them into an earthen vessel with a quart of water and a pinch of salt; let them stew gently for three hours, until the liquid is reduced to half the quantity. Strain through muslin, add a little sugar, and sherry or brandy.

85. Maccaroni Cheese.

Take two ounces of rich new cheese, scrape it very fine, tie it up in muslin, put into cold water (half a pint), let it boil for half an hour, then take the cheese out; put into the liquid half an ounce of best maccaroni with pepper, very little salt, and three or four small pieces of butter; let it simmer gently for an hour and a quarter, then beat up the yolk of an egg with a tablespoonful of cream, stir them into the maccaroni, put into a slightly buttered pie-dish. Place before the fire until it is nicely browned, then serve.

86. Buttered Eggs.

Take one egg, beat it up with one tablespoonful of fresh milk, a little pepper and salt; put a piece of butter the size of a walnut into a small lined saucepan; when it is melted, beat the egg, &c., into it; put it on the fire, beating all the time with a fork, until it is just set; then turn it on to a slice of hot buttered toast, which should be ready before cooking the egg, so that it can be served at once.

87. Gum Water.

Take an ounce of the best gum arabic, two ounces of sugar, put them into an earthen vessel with a pint of water. Let it stand in a saucepan of boiling water, stirring occasionally until dissolved; add the juice of one lemon. This is a useful drink for stopping a tickling cough. It should be taken hot, very little at a time.

88. Meat Lozenges.

Soak an ounce of best isinglass in a pint of beef essence (see recipe No. 1 for beef essence) for one hour; put it into a stewpan; when it boils, skim it continually until no more scum rises, then allow it to boil fast, uncovered, for a short time; when it becomes the consistency of strong gum, pour it on to a dish. When cold cut into lozenges. Put them away in a dry place in a tin box; they will keep for a long time.

89. Raw-meat Sandwiches.

Three ounces of raw beef or mutton, one ounce of very fine bread-crumbs, one teaspoonful of castor-sugar; cut the meat very

fine, rub it through a hair-sieve, then pound it in a mortar into
a paste. Mix with it the bread-crumbs, sugar, a little salt and
pepper; spread it between thin slices of either brown or white
bread and butter. These sandwiches will often create an appetite, if the patient is not allowed to know that the meat is raw.

90. Chicken Soufflée.

Take the flesh of one lightly-boiled chicken; pound it in a
mortar, rub it through a hair-sieve with wooden spoon; mix it
with three gills of good chicken or veal stock, a little pepper and
salt, a little nutmeg (if liked); put it into a clean stewpan, stir
it over the fire for ten minutes. Add the yolks of three eggs, one
at a time (remove the pan off the fire whilst doing so), then add
four stiffly beaten whites (it is best to beat the whites of eggs on
a large dish or flat plate with a bread or carving knife—they
become stiff sooner than any other way); beat lightly until
thoroughly mixed, then pour into buttered soufflée case. Bake
in a hot oven for fifteen or twenty minutes. Serve at once.

91. Beef or Mutton Tea for Anæmic Patients.

One pound of lean beef, or one and a half pound of lean mutton,
as freshly killed as possible. Chop the meat *very* fine, add to it
one pint of cold water, ten drops of hydrochloric acid, and let it
stand for six hours. Put it into a stewpan over a slow fire, allow
it just to boil, then draw it to the side, and let it simmer gently
for two hours; strain it through a rather coarse sieve; remove
all fat with blotting-paper at once. Add a little pepper and salt.
The patient should take about a port-wine glassful twice daily.

92. Tapioca Jelly.

Take a teacupful of best tapioca, soak it in one pint of cold
water for six hours, then put it into a saucepan with a little salt,
one tablespoonful of sifted sugar, and the rind and juice of one
lemon; add another pint of water; stir until it boils. Let it boil
for twenty minutes; take the rind of the lemon out, add a claret
glassful of either port or sherry, or a port-wine glassful of brandy.
Whilst cooling turn into a mould. When cold it is ready for use.

93. Savoury Custard.

One gill of hot chicken-broth (see recipe for chicken-broth, No. 8), a little pepper and salt (a little chopped parsley is a great improvement if allowed). Take two eggs, beat well; add the hot broth, salt, pepper, &c., stirring carefully all the time. Pour it into a buttered mould, then place it in a saucepan of boiling water up to the same height the custard is in the mould (the mould must be covered with buttered paper). Steam it gently for thirty to forty minutes. Turn out on to a hot dish, served with gravy poured round.

94. Nutritious Chicken-Broth.

One fine young chicken (not fat), two pounds of the scrag end of neck of veal, salt, pepper, celery, and any herbs which may be allowed (for a delicate digestion only a little salt and pepper). Cut both chicken and veal into small pieces, also putting the bones of the chicken with the meat in two quarts of water into an earthen vessel. Let it stew very gently in the oven for five hours. When cold take off all fat.

Chicken-broth, as well as beef-tea, mutton-broth, or any soup, is more digestible for patients, thickened with either arrowroot or boiled flour. Barley or rice is sometimes ordered to thicken broth; few patients like them. Bread-crumbs rubbed through a hair-sieve into the broth, and allowed to simmer for five minutes, thicken it nicely.

95. Lamb's Foot.

Soak one lamb's foot in boiling water for half an hour, then put it in half a pint of veal stock, salt and pepper to taste (any herbs if allowed), into a clean stewpan. Let it simmer gently for an hour and a half, then serve with white sauce as follows:— Half an ounce of butter, half an ounce of flour, half a gill of the stock in which the foot was stewed, half a gill of cream, half a tea-spoonful of chopped parsley, and a few drops of lemon-juice mixed. Mix the flour smoothly into a thin paste with the stock, add the butter and cream, stir gently over the fire for a few minutes, add the other ingredients gradually. Let all simmer gently together for five minutes, stirring all the time.

96. Calf's Foot.

Soak one calf's foot in three pints of water all night, pour the water over it boiling, then let it remain until the next morning. Put it into a clean stewpan; let it simmer for three hours in the water it has stood in, add salt and pepper. Remove all the large bones before serving. Serve with sauce as for lamb's foot, only of course made with double the quantity.

97. Milk and Oatmeal.

Put a breakfastcupful of milk into a clean saucepan; when it begins to simmer, sprinkle into it a small tablespoonful of fine oatmeal, stirring quickly all the time. Let it simmer for half an hour, add a little salt or sugar according to taste.

INDEX.

PRINTED BY BALLANTYNE, HANSON AND CO.
EDINBURGH AND LONDON.

A Classified Catalogue of Books on Medicine and the Collateral Sciences, Pharmacy, Dentistry, Chemistry, Hygiene, Microscopy, Etc.

P. Blakiston's Son & Company, Publishers of Medical and Scientific Books, 1012 Walnut Street, Philadelphia

No. 8. 6–3–01.

SUBJECT INDEX.

Special Catalogues of Books on Pharmacy, Dentistry, Chemistry, Hygiene, and Nursing will be sent free upon application. All inquiries regarding prices, dates of edition, terms, etc., will receive prompt attention.

Self-Examination for Medical Students. 3500 Questions on Medical Subjects, with References to Standard Works in which the correct replies will be found Together with Questions from State Examining Boards. 3d Edition. *Just Ready*. Paper Cover, 10 cts.

SPECIAL NOTE.—The prices given in this catalogue are net, no discount can be allowed retail purchasers under any consideration. This rule has been established in order that everyone will be treated alike, a general reduction in former prices having been made to meet previous retail discounts. Upon receipt of the advertised price any book will be forwarded by mail or express, all charges prepaid.

ANATOMY.

MORRIS. Text-Book of Anatomy. 2d Edition. Revised and Enlarged. 790 Illustrations, 214 of which are printed in colors. *Thumb Index in Each Copy.* Cloth, $6.00; Leather, $7.00
" The ever-growing popularity of the book with teachers and students is an index of its value."—*Medical Record, New York.*

BROOMELL. Anatomy and Histology of the Human Mouth and Teeth. 284 Illustrations. $4.50

CAMPBELL. Dissection Outlines. Based on Morris' Anatomy. 2d Edition. .50

DEAVER. Surgical Anatomy. A Treatise on Anatomy in its Application to Medicine and Surgery. With 400 very Handsome full-page Illustrations Engraved from Original Drawings made by special Artists from dissections prepared for the purpose. Three Volumes. Cloth, $21.00; Half Morocco or Sheep, $24.00; Half Russia, $27.00

GORDINIER. Anatomy of the Central Nervous System. With 271 Illustrations, many of which are original. Cloth, $6.00

HEATH. Practical Anatomy. 8th Edition. 300 Illus. $4.25

HOLDEN. Anatomy. A Manual of Dissections. Revised by A. HEWSON, M.D., Demonstrator of Anatomy, Jefferson Medical College, Philadelphia. Over 300 handsome Illustrations. 7th Edition. In two compact 12mo Volumes. Large New Type. *Just Ready.*
Vol. I. Scalp—Face—Orbit—Neck—Thorax—Upper Extremity.
$1.50
Vol. II. Abdomen—Perineum—Lower Extremity—Brain—Eye—Ear—Mammary Gland—Scrotum—Testes. $1.50

HOLDEN. Human Osteology. Comprising a Description of the Bones, with Colored Delineations of the Attachments of the Muscles. The General and Microscopical Structure of Bone and its Development. With Lithographic Plates and numerous Illus. 8th Ed. $5.25

HOLDEN. Landmarks. Medical and Surgical. 4th Ed. .75

HUGHES AND KEITH. Dissections. In three Parts: I, Upper and Lower Extremity; II, Abdomen, Pelvis; III, Perineum, Thorax. With Colored and other Illustrations. *In Press.*

MACALISTER. Human Anatomy. Systematic and Topographical. With Special Reference to the Requirements of Practical Surgery and Medicine. 816 Illustrations. Cloth, $5.00; Leather, $6.00

MARSHALL. Physiological Diagrams. Life Size. Colored. Eleven Life-Size Diagrams (each seven feet by three feet seven inches). Designed for Demonstration before the Class.
In Sheets, Unmounted, $40.00; Backed with Muslin and Mounted on Rollers, $60.00; Ditto, Spring Rollers, in Handsome Walnut Wall Map Case, $100.00; Single Plates—Sheets, $5.00; Mounted, $7.50. Explanatory Key, .50. *Purchaser must pay freight charges.*

POTTER. Compend of Anatomy, Including Visceral Anatomy. 6th Ed. 16 Lith. Plates and 117 other Illus. .80; Interleaved, $1.00

WILSON. Anatomy. 11th Edition. 429 Illus., 26 Plates. $5.00

WINDLE. Surface Anatomy. Colored and other Illus. $1.00

BRAIN AND INSANITY (see also Nervous Diseases).

BLACKBURN. A Manual of Autopsies. Designed for the Use of Hospitals for the Insane and other Public Institutions. Ten full-page Plates and other Illustrations. $1.25

DERCUM. Mental Therapeutics, Rest, etc.   *Nearly Ready.*

GORDINIER. The Gross and Minute Anatomy of the Central Nervous System. With full-page and other Illustrations. $6.00

HORSLEY. The Brain and Spinal Cord. The Structure and Functions of. Numerous Illustrations. $2.50

IRELAND. The Mental Affections of Children. 2d Ed. $4 00

LEWIS (BEVAN). Mental Diseases. A Text-Book Having Special Reference to the Pathological Aspects of Insanity. 26 Lithographic Plates and other Illustrations. 2d Ed.   *Just Ready.* $7.00

MANN. Manual of Psychological Medicine and Allied Nervous Diseases. $3.00

PERSHING. Diagnosis of Nervous and Mental Disease. Illustrated.   *In Press.*

REGIS. Mental Medicine. Authorized Translation by H. M. BANNISTER, M.D. $2.00

SHUTTLEWORTH. Mentally Deficient Children. $1.50

STEARNS. Mental Diseases. With a Digest of Laws Relating to Care of Insane. Illustrated.   Cloth, $2.75; Sheep, $3.25

TUKE. Dictionary of Psychological Medicine. Giving the Definition, Etymology, and Symptoms of the Terms used in Medical Psychology, with the Symptoms, Pathology, and Treatment of the Recognized Forms of Mental Disorders. Two volumes. $10.00

WOOD, H. C. Brain and Overwork.   .40

CHEMISTRY AND TECHNOLOGY.

Special Catalogue of Chemical Books sent free upon application.

ALLEN. Commercial Organic Analysis. A Treatise on the Modes of Assaying the Various Organic Chemicals and Products Employed in the Arts, Manufactures, Medicine, etc., with concise methods for the Detection of Impurities, Adulterations, etc. 8vo.

Vol. I. Alcohols, Neutral Alcoholic Derivatives, etc., Ethers, Vegetable Acids, Starch, Sugars, etc. 3d Edition, by HENRY LEFFMANN, M. D. $4.50

Vol. II, Part I. Fixed Oils and Fats, Glycerol, Explosives, etc. 3d Edition, by HENRY LEFFMANN, M. D. $3.50

Vol. II, Part II. Hydrocarbons, Mineral Oils, Lubricants, Benzenes, Naphthalenes and Derivatives, Creosote, Phenols, etc. 3d Edition, by HENRY LEFFMANN, M.D. $3.50

Vol. II, Part III. Terpenes, Essential Oils, Resins, Camphors, etc. 3d Edition, by HENRY LEFFMANN, M.D.   *Preparing.*

Vol. III, Part I. Tannins, Dyes and Coloring Matters. 3d Edition. Enlarged and Rewritten by J. MERRITT MATTHEWS, PH.D., of the Philadelphia Textile School. Illustrated. $4.10

Vol. III, Part II. The Amines, Hydrazines and Derivatives, Pyridine Bases. The Antipyretics, etc. Vegetable Alkaloids, Tea, Coffee, Cocoa, etc. 8vo. 2d Edition. $4 50

Vol. III, Part III. Vegetable Alkaloids, Non-Basic Vegetable Bitter Principles. Animal Bases, Animal Acids, Cyanogen Compounds, etc. 2d Edition, 8vo. $4.50

Vol. IV. The Proteids and Albuminous Principles. 2d Ed. $4.50

ALLEN. Albuminous and Diabetic Urine. Illustrated. $2.25

BARTLEY. Medical and Pharmaceutical Chemistry. A Text-Book for Medical, Dental, and Pharmaceutical Students. With Illustrations, Glossary, and Complete Index. 5th Edition. $3.00

BARTLEY. Clinical Chemistry. The Examination of Feces, Saliva, Gastric Juice, Milk, and Urine. $1.00

BLOXAM. Chemistry, Inorganic and Organic. With Experiments. 8th Ed., Revised. 281 Engravings. Clo., $4.25; Lea., $5.25

CALDWELL. Elements of Qualitative and Quantitative Chemical Analysis. 3d Edition, Revised. $1.00

CAMERON. Oils and Varnishes. With Illustrations. $2.25

CAMERON. Soap and Candles. 54 Illustrations. $2.00

CLOWES AND COLEMAN. Quantitative Analysis. 5th Edition. 122 Illustrations. $3.50

COBLENTZ. Volumetric Analysis. Illustrated. *In Press.*

CONGDON. Laboratory Instructions in Chemistry. With Numerous Tables and 56 Illustrations. *Just Ready.* $1.00

GARDNER. The Brewer, Distiller, and Wine Manufacturer. Illustrated. $1.50

GRAY. Physics. Volume I. Dynamics and Properties of Matter. 350 Illustrations. *Just Ready.* $4.50

GROVES AND THORP. Chemical Technology. The Application of Chemistry to the Arts and Manufactures.
Vol. I. Fuel and Its Applications. 607 Illustrations and 4 Plates.
Cloth, $5.00; ½ Mor., $6.50
Vol. II. Lighting. Illustrated. Cloth, $4.00; ½ Mor., $5.50
Vol. III. Gas Lighting. Cloth, $3.50; ½ Mor., $4.50
Vol. IV. Electric Lighting. Photometry. *In Press.*

HOLLAND. The Urine, the Gastric Contents, the Common Poisons, and the Milk. Memoranda, Chemical and Microscopical, for Laboratory Use. 6th Ed. Illustrated and interleaved, $1.00

LEFFMANN. Compend of Medical Chemistry, Inorganic and Organic. 4th Edition, Revised. .80; Interleaved, $1.00

LEFFMANN. Analysis of Milk and Milk Products. 2d Edition, Enlarged. Illustrated. $1.25

LEFFMANN. Water Analysis. For Sanitary and Technic Purposes. Illustrated. 4th Edition. $1.25

LEFFMANN. Structural Formulæ. Including 180 Structural and Stereo-Chemical Formulæ. 12mo. Interleaved. $1.00

LEFFMANN AND BEAM. Select Methods in Food Analysis. Illustrated *Just Ready.* $2 50

MUTER. Practical and Analytical Chemistry. 2d American from the Eighth English Edition. Revised to meet the requirements of American Students. 56 Illustrations. $1.25

OETTEL. Exercises in Electro-Chemistry. Illustrated. .75

OETTEL. Electro-Chemical Experiments. Illustrated. .75

RICHTER. Inorganic Chemistry. 5th American from 10th German Edition. Authorized translation by EDGAR F. SMITH, M.A., PH.D. 89 Illustrations and a Colored Plate. $1.75

RICHTER. Organic Chemistry. 3d American Edition. Trans. from the 8th German by EDGAR F. SMITH. Illustrated. 2 Volumes.
Vol. I. Aliphatic Series. 625 Pages. $3.00
Vol. II. Carbocyclic Series. 671 Pages. $3.00

SMITH. Electro-Chemical Analysis. 2d Ed. 28 Illus. $1.25

SMITH AND KELLER. Experiments. Arranged for Students in General Chemistry. 4th Edition. Illustrated. .60

SUTTON. Volumetric Analysis. A Systematic Handbook for the Quantitative Estimation of Chemical Substances by Measure, Applied to Liquids, Solids, and Gases. 8th Edition, Revised. 112 Illustrations. $5.00

SYMONDS. Manual of Chemistry, for Medical Students. 2d Edition. $2.00

TRAUBE. Physico-Chemical Methods. Translated by Hardin. 97 Illustrations. $1.50

THRESH. Water and Water Supplies. 2d Edition. $2.00

ULZER AND FRAENKEL. Chemical Technical Analysis. Translated by Fleck. Illustrated. $1.25

WOODY. Essentials of Chemistry and Urinalysis. 4th Edition. Illustrated. $1.50

⁎ *Special Catalogue of Books on Chemistry free upon application.*

CHILDREN.

CAUTLEY. Feeding of Infants and Young Children by Natural and Artificial Methods. $2.00

HALE. On the Management of Children. .50

HATFIELD. Compend of Diseases of Children. With a Colored Plate. 2d Edition. .80 ; Interleaved, $1.00

IRELAND. The Mental Affections of Children. 2d Ed. $4.00

MEIGS. Infant Feeding and Milk Analysis. The Examination of Human and Cow's Milk, Cream, Condensed Milk, etc., and Directions as to the Diet of Young Infants. .50

POWER. Surgical Diseases of Children and their Treatment by Modern Methods. Illustrated. $2.50

SHUTTLEWORTH. Mentally Deficient Children. New Edition. $1.50

STARR. The Digestive Organs in Childhood. The Diseases of the Digestive Organs in Infancy and Childhood. With Chapters on the Investigation of Disease and the Management of Children. 3d Edition, Rewritten and Enlarged. Illustrated by two Colored Plates and numerous Wood Engravings. *In Preparation.*

STARR. Hygiene of the Nursery. Including the General Regimen and Feeding of Infants and Children, and the Domestic Management of the Ordinary Emergencies of Early Life, Massage, etc. 6th Edition. 25 Illustrations. $1.00

SMITH. Wasting Diseases of Children. 6th Edition. $2.00

TAYLOR AND WELLS. The Diseases of Children. 2d Edition, Revised and Enlarged. Illustrated. 8vo. *Just Ready.* $4.50

CLINICAL CHARTS.

GRIFFITH. Graphic Clinical Chart for Recording Temperature, Respiration, Pulse, Day of Disease, Date, Age, Sex, Occupation, Name, etc. Printed in three colors. Sample copies free. Put up in loose packages of fifty, .50. Price to Hospitals, 500 copies, $4.00 ; 1000 copies, $7.50. With name of Hospital printed on, .50 extra.

KEEN'S CLINICAL CHARTS. Seven Outline Drawings of the Body, on which may be marked the Course of Disease, Fractures, Operations, etc. Pads of fifty, $1.00. Each Drawing may also be had separately, twenty-five to pad, 25 cents.

SCHREINER. Diet Lists. Arranged in the form of a chart. With Pamphlets of Specimen Dietaries. Pads of 50.   .75

DENTISTRY.

Special Catalogue of Dental Books sent free upon application.

BARRETT. Dental Surgery for General Practitioners and Students of Medicine and Dentistry. Extraction of Teeth, etc. 3d Edition. Illustrated.   *Nearly Ready.*

BROOMELL. Anatomy and Histology of the Human Mouth and Teeth. 284 Handsome Illustrations. $4.50

FILLEBROWN. A Text-Book of Operative Dentistry. Written by invitation of the National Association of Dental Faculties. Illustrated. $2.25

GORGAS. Dental Medicine. A Manual of Materia Medica and Therapeutics. 7th Edition, Revised.   Cloth, $4.00; Sheep, $5.00

GORGAS. Questions and Answers for the Dental Student. Embracing all the subjects in the Curriculum of the Dental Student. Octavo. *Just Ready.* $6.00

HARRIS. Principles and Practice of Dentistry. Including Anatomy, Physiology, Pathology, Therapeutics, Dental Surgery, and Mechanism. 13th Edition. Revised by F. J. S. GORGAS, M.D., D.D.S. 1250 Illustrations.   Cloth, $6.00; Leather, $7.00

HARRIS. Dictionary of Dentistry. Including Definitions of Such Words and Phrases of the Collateral Sciences as Pertain to the Art and Practice of Dentistry. 6th Edition. Revised and Enlarged by FERDINAND F. S. GORGAS, M.D., D.D.S.   Cloth, $5.00; Leather, $6.00

HEATH. Injuries and Diseases of the Jaws. 4th Edition. 187 Illustrations. $4.50

RICHARDSON. Mechanical Dentistry. 7th Edition. Thoroughly Revised and Enlarged by DR. GEO. W. WARREN. 691 Illustrations.   Cloth, $5.00; Leather, $6.00

SMITH. Dental Metallurgy. Illustrated. $1.75

TAFT. Index of Dental Periodical Literature. $2.00

TOMES. Dental Anatomy. Human and Comparative. 263 Illustrations. 5th Edition. $4.00

TOMES. Dental Surgery. 4th Edition. 289 Illustrations. $4.00

WARREN. Compend of Dental Pathology and Dental Medicine. With a Chapter on Emergencies. 3d Edition. Illustrated.
.80; Interleaved, $1.25

WARREN. Dental Prosthesis and Metallurgy. 129 Ills. $1.25

WHITE. The Mouth and Teeth. Illustrated.   .40

DICTIONARIES.

GOULD. **The Illustrated Dictionary of Medicine, Biology, and Allied Sciences.** Being an Exhaustive Lexicon of Medicine and those Sciences Collateral to it: Biology (Zoology and Botany), Chemistry, Dentistry, Parmacology, Microscopy, etc., with many useful Tables and numerous fine Illustrations. 1633 pages. 5th Ed.
Sheep or Half Dark Green Leather, $10.00; Thumb Index, $11.00
Half Russia, Thumb Index, $12.00

GOULD. **The Medical Student's Dictionary.** 11th Edition. Illustrated. Including all the Words and Phrases Generally Used in Medicine, with their Proper Pronunciation and Definition, Based on Recent Medical Literature. With a new Table of Eponymic Terms and Tests and Tables of the Bacilli, Micrococci, Mineral Springs, etc., of the Arteries, Muscles, Nerves, Ganglia, Plexuses, etc. 11th Edition. Enlarged by over 100 pages and illustrated with a large number of engravings. 840 pages.
Half Green Morocco, $2.50; Thumb Index, $3.00

GOULD. **The Pocket Pronouncing Medical Lexicon.** 4th Edition. (30,000 Medical Words Pronounced and Defined.) Containing all the Words, their Definition and Pronunciation, that the Medical, Dental, or Pharmaceutical Student Generally Comes in Contact With; also Elaborate Tables of Eponymic Terms. Arteries, Muscles, Nerves, Bacilli, etc., etc., a Dose List in both English and Metric Systems, etc., Arranged in a Most Convenient Form for Reference and Memorizing. A new (Fourth) Edition, Revised and Enlarged. 838 pages.
Full Limp Leather, Gilt Edges, $1.00; Thumb Index, $1.25
120,000 Copies of Gould's Dictionaries Have Been Sold.

GOULD AND PYLE. **Cyclopedia of Practical Medicine and Surgery.** Seventy-two Special Contributors. Illustrated. One Volume. A Concise Reference Handbook, Alphabetically Arranged, of Medicine, Surgery, Obstetrics, Materia Medica, Therapeutics, and the Various Specialties, with Particular Reference to Diagnosis and Treatment. Compiled under the Editorial Supervision of GEORGE M. GOULD, M.D., Author of "An Illustrated Dictionary of Medicine" · Editor "Philadelphia Medical Journal," etc.; and WALTER L. PYLE, M.D., Assistant Surgeon Wills Eye Hospital ; formerly Editor "International Medical Magazine," etc., and Seventy-two Special Contributors. With many Illustrations. Large Square 8vo. to correspond with Gould's "Illustrated Dictionary." *Just Ready.* Full Sheep or Half Dark-Green Leather, $10.00 With Thumb Index, $11.00; Half Russia, Thumb Index, $12.00 net.

₀ Sample Pages and Illustrations and Descriptive Circulars of Gould's Dictionaries and Cyclopedia sent free upon application.

HARRIS. **Dictionary of Dentistry.** Including Definitions of Such Words and Phrases of the Collateral Sciences as Pertain to the Art and Practice of Dentistry. 6th Edition. Revised and Enlarged by FERDINAND J. S. GORGAS, M.D., D.D.S. Cloth, $5.00; Leather, $6.00

LONGLEY. **Pocket Medical Dictionary.** With an Appendix, containing Poisons and their Antidotes, Abbreviations used in Prescriptions, etc. Cloth, .75; Tucks and Pocket, $1.00

MAXWELL. **Terminologia Medica Polyglotta.** By Dr. THEODORE MAXWELL, Assisted by Others. $3.00
The object of this work is to assist the medical men of any nationality in reading medical literature written in a language not their own. Each term is usually given in seven languages, viz.: English, French, German, Italian, Spanish, Russian, and Latin.

TREVES AND LANG. **German-English Medical Dictionary.**
Half Russia, $3.25

EAR (see also Throat and Nose).

BURNETT. Hearing and How to Keep It. Illustrated. .40

DALBY. Diseases and Injuries of the Ear. 4th Edition. 38 Wood Engravings and 8 Colored Plates. $2.50

HOVELL. Diseases of the Ear and Naso-Pharynx. Including Anatomy and Physiology of the Organ, together with the Treatment of the Affections of the Nose and Pharynx which Conduce to Aural·Disease. 128 Illustrations. 2d Edition. *Just Ready.* $5.50

PRITCHARD. Diseases of the Ear. 3d Edition, Enlarged. Many Illustrations and Formulæ. $1.50

WOAKES. Deafness, Giddiness, and Noises in the Head. 4th Edition. Illustrated. $2.00

ELECTRICITY.

BIGELOW. Plain Talks on Medical Electricity and Batteries. With a Therapeutic Index and a Glossary. 43 Illustrations. 2d Edition. $1.00

HEDLEY. Therapeutic Electricity and Practical Muscle Testing. 99 Illustrations. $2.50

JACOBY. Electrotherapy. 2 Volumes. Illustrated. Including Special Articles by Special Authors. *Just Ready.*

JONES. Medical Electricity. 3d Edition. 117 Illus. $3 00

EYE.

A Special Circular of Books on the Eye sent free upon application.

DONDERS. The Nature and Consequences of Anomalies of Refraction. With Portrait and Illustrations. Half Morocco, $1.25

FICK. Diseases of the Eye and Ophthalmoscopy. Translated by A. B. HALE, M. D. 157 Illustrations, many of which are in colors, and a glossary. Cloth, $4.50; Sheep, $5.50

GOULD AND PYLE. Compend of Diseases of the Eye and Refraction. Including Treatment and Operations, and a Section on Local Therapeutics. With Formulæ, Useful Tables, a Glossary, and 111 Illus., several of which are in colors. 2d Edition, Revised. Cloth, .80; Interleaved, $1.00

GOWERS. Medical Ophthalmoscopy. A Manual and Atlas with Colored Autotype and Lithographic Plates and Wood-cuts, Comprising Original Illustrations of the Changes of the Eye in Diseases of the Brain, Kidney, etc. 3d Edition. $4.00

HARLAN. Eyesight, and How to Care for It. Illus. .40

HARTRIDGE. Refraction. 104 Illustrations and Test Types. 10th Edition, Enlarged. $1.50

HARTRIDGE. On the Ophthalmoscope. 3d Edition. With 4 Colored Plates and 68 Wood-cuts. $1.50

HANSELL AND REBER. Muscular Anomalies of the Eye. Illustrated. $1.50

HANSELL AND BELL. Clinical Ophthalmology. Colored Plate of Normal Fundus and 120 Illustrations. $1.50

JESSOP. Manual of Ophthalmic Surgery and Medicine. Colored Plates and 108 other Illustrations. Cloth, $3.00

MORTON. Refraction of the Eye. Its Diagnosis and the Correction of its Errors. 6th Edition. $1.00

OHLEMANN. Ocular Therapeutics. Authorized Translation, and Edited by DR. CHARLES A. OLIVER. $1.75

PHILLIPS. Spectacles and Eyeglasses. Their Prescription and Adjustment. 2d Edition. 49 Illustrations. $1.00

SWANZY. Diseases of the Eye and Their Treatment. 7th Edition, Revised and Enlarged. 164 Illustrations, 1 Plain Plate, and a Zephyr Test Card. $2.50

THORINGTON. Retinoscopy. 4th Edition. Carefully Revised. Illustrated. *Just Ready.* $1.00

THORINGTON. Refraction and How to Refract. 200 Illustrations, 13 of which are Colored. 2d Edition. $1.50

WALKER. Students' Aid in Ophthalmology. Colored Plate and 40 other Illustrations and Glossary. $1.50

WRIGHT. Ophthalmology. 2d Edition, Revised and Enlarged. 117 Illustrations and a Glossary. *Just Ready.* $3.00

FEVERS.

GOODALL AND WASHBOURN. Fevers and Their Treatment. Illustrated. $3.00

GOUT AND RHEUMATISM.

DUCKWORTH. A Treatise on Gout. With Chromo-lithographs and Engravings.   Cloth, $6.00

HAIG. Causation of Disease by Uric Acid. A Contribution to the Pathology of High Arterial Tension. Headache, Epilepsy, Gout, Rheumatism, Diabetes, etc. 5th Edition. *Just Ready.* $3 00

HEART.

THORNE. The Schott Methods of the Treatment of Chronic Heart Disease. Third Edition. Illustrated. $1 75

HISTOLOGY.

CUSHING. Compend of Histology. Illustrated.
          Nearly Ready. .80; Interleaved, $1.00

STIRLING. Outlines of Practical Histology. 368 Illustrations. 2d Edition. Revised and Enlarged. With new Illustrations. $2.00

STÖHR. Histology and Microscopical Anatomy. Edited by A. SCHAPER, M.D., University of Breslau, formerly Demonstrator of Histology, Harvard Medical School. Third American from 8th German Edition, Revised and Enlarged. 301 Illus. *Just Ready.* $3.00

HYGIENE AND WATER ANALYSIS.

Special Catalogue of Books on Hygiene sent free upon application.

CANFIELD. Hygiene of the Sick-Room. A Book for Nurses and Others. Being a Brief Consideration of Asepsis, Antisepsis, Disinfection, Bacteriology, Immunity, Heating, Ventilation, etc. $1.25

COPLIN. Practical Hygiene. A Complete American Text-Book. 138 Illustrations. New Edition. *Preparing.*

ERNST AND ABRAMS. Prophylaxis and Personal Hygiene. *In Press.*

HARTSHORNE. Our Homes. Illustrated. .40

KENWOOD. Public Health Laboratory Work. 116 Illustrations and 3 Plates. $2.00

LEFFMANN. Examination of Water for Sanitary and Technical Purposes. 4th Edition. Illustrated. $1.25

LEFFMANN. Analysis of Milk and Milk Products. Illustrated. Second Edition. $1.25

LINCOLN. School and Industrial Hygiene. .40

McNEILL. The Prevention of Epidemics and the Construction and Management of Isolation Hospitals. Numerous Plans and Illustrations. $3.50

NOTTER. The Theory and Practice of Hygiene. 15 Plates and 138 other Illustrations. 8vo. 2d Edition. $7.00

PARKES. Hygiene and Public Health. By Louis C. Parkes, M.D. 6th Edition. Enlarged. Illustrated. *Just Ready.* $3.00

PARKES. Popular Hygiene. The Elements of Health. A Book for Lay Readers. Illustrated. $1.25

STARR. The Hygiene of the Nursery. Including the General Regimen and Feeding of Infants and Children, and the Domestic Management of the Ordinary Emergencies of Early Life, Massage, etc. 6th Edition. 25 Illustrations. $1.00

STEVENSON AND MURPHY. A Treatise on Hygiene. By Various Authors. In Three Octave Volumes. Illustrated.
Vol. I, $6.00; Vol. II, $6.00; Vol. III, $5.00

₊ Each Volume sold separately. Special Circular upon application.

THRESH. Water and Water Supplies. 2d Edition. $2.00

WILSON. Hand-Book of Hygiene and Sanitary Science. With Illustrations. 8th Edition. $3.00

WEYL. Sanitary Relations of the Coal-Tar Colors. Authorized Translation by HENRY LEFFMANN, M.D., PH.D. $1.25

LUNGS AND PLEURÆ.

HARRIS AND BEALE. Treatment of Pulmonary Consumption. $2.50

KNOPF. Pulmonary Tuberculosis. Its Modern Prophylaxis and Treatment in Special Institutions and at Home. Illus. $3.00

STEEL. Physical Signs of Pulmonary Disease. Illus. $1.25

MASSAGE—PHYSICAL EXERCISE.

KLEEN. Hand-Book of Massage. Authorized translation by MUSSEY HARTWELL, M.D., PH.D. With an Introduction by Dr. S. WEIR MITCHELL. Illustrated by a series of Photographs Made Especially by DR. KLEEN for the American Edition. $2.25

OSTROM. Massage and the Original Swedish Movements. Their Application to Various Diseases of the Body. A Manual for Students, Nurses, and Physicians. Fourth Edition, Enlarged. 105 Illustrations, many of which are original. $1.00

MITCHELL AND GULICK. Mechanotherapy. Illus. *In Press.*

TREVES. Physical Education. Methods, etc.   .75

WARD. Notes on Massage. Interleaved.   Paper cover, $1.00

MATERIA MEDICA AND THERAPEUTICS.

BIDDLE. Materia Medica and Therapeutics. Including Dose List, Dietary for the Sick, Table of Parasites, and Memoranda of New Remedies. 13th Edition, Revised. 64 Illustrations and a Clinical Index.   Cloth, $4.00; Sheep, $5.00

BRACKEN. Outlines of Materia Medica and Pharmacology. $2.75

COBLENTZ. The Newer Remedies. Including their Synonyms, Sources, Methods of Preparation, Tests, Solubilities, Doses, etc. 3d Edition, Enlarged and Revised. $1.00

COHEN. Physiologic Therapeutics. Mechanotherapy, Mental Therapeutics, Electrotherapy. Climatology, Hydrotherapy, Pneumatotherapy, Prophylaxis, Dietetics, etc. 11 Volumes, Octavo. Illustrated.   Price for the set, $22.00

Special Descriptive Circular will be sent upon application.

DAVIS. Materia Medica and Prescription Writing. $1.50

GORGAS. Dental Medicine. A Manual of Materia Medica and Therapeutics. 7th Edition, Revised. $4.00

GROFF. Materia Medica for Nurses, with questions for Self Examination and a complete Glossary. $1.25

HELLER. Essentials of Materia Medica, Pharmacy, and Prescription Writing. $1.50

MAYS. Theine in the Treatment of Neuralgia. ½ bound, .50

POTTER. Hand-Book of Materia Medica, Pharmacy, and Therapeutics, including the Action of Medicines, Special Therapeutics, Pharmacology, etc., including over 600 Prescriptions and Formulæ. 8th Edition. Revised and Enlarged. With Thumb Index in each copy. *Just Ready.*   Cloth, $5.00; Sheep, $6.co

POTTER. Compend of Materia Medica, Therapeutics, and Prescription Writing, with Special Reference to the Physiological Action of Drugs. 6th Edition.   .80: Interleaved, $1.00

MURRAY. Rough Notes on Remedies. 4th Edition. $1.25

SAYRE. Organic Materia Medica and Pharmacognosy. An Introduction to the Study of the Vegetable Kingdom and the Vegetable and Animal Drugs. Comprising the Botanical and Physical Characteristics, Source, Constituents, and Pharmacopeial Preparations, Insects Injurious to Drugs, and Pharmacal Botany. With sections on Histology and Microtechnique, by W. C. STEVENS. 374 Illustrations, many of which are original. 2d Edition.
Cloth, $4.50

WHITE AND WILCOX. Materia Medica, Pharmacy, Pharmacology, and Therapeutics. 4th American Edition, Revised by REYNOLD W. WILCOX, M.A., M.D., LL.D., Professor of Clinical Medicine and Therapeutics at the New York Post-Graduate Medical School.
Cloth, $3.00; Leather, $3.50

" The care with which Dr. Wilcox has performed his work is conspicuous on every page, and it is evident that no recent drug possessing any merit has escaped his eye. We believe, on the whole, this is the best book on Materia Medica and Therapeutics to place in the hands of students, and the practitioner will find it a most satisfactory work for daily use."—*The Cleveland Medical Gazette.*

MEDICAL JURISPRUDENCE AND TOXICOLOGY.

REESE. Medical Jurisprudence and Toxicology. A Text-Book for Medical and Legal Practitioners and Students. 5th Edition. Revised by HENRY LEFFMANN, M.D. Clo., $3.00; Leather, $3.50

" To the student of medical jurisprudence and toxicology it is invaluable, as it is concise, clear, and thorough in every respect."—*The American Journal of the Medical Sciences.*

MANN. Forensic Medicine and Toxicology. Illus. $6.50

TANNER. Memoranda of Poisons. Their Antidotes and Tests. 8th Edition, by DR HENRY LEFFMANN. *Just Ready.*   .75

MICROSCOPY.

CARPENTER. The Microscope and Its Revelations. 8th Edition, Revised and Enlarged. Over 800 Illustrations and many Lithographs.   *Just Ready.*

LEE. The Microtomist's Vade Mecum. A Hand-Book of Methods of Microscopical Anatomy. 887 Articles. 5th Edition, Enlarged. $4.00

REEVES. Medical Microscopy, including Chapters on Bacteriology, Neoplasms, Urinary Examination, etc. Numerous Illustrations, some of which are printed in colors. $2.50

WETHERED. Medical Microscopy. A Guide to the Use of the Microscope in Practical Medicine. 100 Illustrations. $2.00

MISCELLANEOUS.

BERRY. Diseases of Thyroid Gland. Illustrated. *In Press.*
BURNETT. Foods and Dietaries. A Manual of Clinical Dietetics. 2d Edition. $1.50
BUXTON. Anesthetics. Illustrated. 3d Edition. $1.50
COHEN. Organotherapy. *In Press.*
DAVIS. Dietotherapy. Food in Health and Disease. *In Press.*
FENWICK. Ulcer of the Stomach. 42 Illustrations. $3.50
GOULD. Borderland Studies. Miscellaneous Addresses and Essays. 12mo. $2.00
GREENE. Medical Examination for Life Insurance. Illustrated. $4.00
HAIG. Causation of Disease by Uric Acid. The Pathology of High Arterial Tension, Headache, Epilepsy, Gout, Rheumatism, Diabetes, Bright's Disease, etc. 5th Edition. $3.00
HAIG. Diet and Food. Considered in Relation to Strength and Power of Endurance. 3d Edition. *Just Ready.* $1.00
HEMMETER. Diseases of the Stomach. Their Special Pathology, Diagnosis, and Treatment. With Sections on Anatomy, Dietetics, Surgery, etc. 2d Edition, Revised and Enlarged. Illustrated. Cloth, $6.00; Sheep, $7.00
HEMMETER. Diseases of the Intestines. Illustrated. 2 Volumes. 8vo. *Nearly Ready.*
HENRY. A Practical Treatise on Anemia. Half Cloth, .50
LEFFMANN. Food Analysis. Illustrated. *Just Ready.* $2.50
NEW SYDENHAM SOCIETY'S PUBLICATIONS. Circulars upon application. Per Annum, $8.00
OSGOOD. The Winter and Its Dangers. .40
OSLER AND McCRAE. Cancer of the Stomach. $2.00
PACKARD. Sea Air and Sea Bathing. .40
RICHARDSON. Long Life and How to Reach It. .40
ST. CLAIR. Medical Latin. $1.00
TISSIER. Pneumatotherapy. *In Press.*
TURNBULL. Artificial Anesthesia. 4th Edition. Illus. $2.50
WEBER AND HINSDALE. Climatology. *Just Ready.*
WILSON. The Summer and Its Diseases. .40
WINTERNITZ. Hydrotherapy. Illustrated. *In Press.*

NERVOUS DISEASES.

BEEVOR. Diseases of the Nervous System and their Treatment. $2.50
DERCUM. Rest, Hypnotism, Mental Therapeutics. *In Press.*
GORDINIER. The Gross and Minute Anatomy of the Central Nervous System. With 271 original Colored and other Illustrations. Cloth, $6.00; Sheep, $7.00
GOWERS. Manual of Diseases of the Nervous System. A Complete Text-Book. Revised, Enlarged, and in many parts Rewritten. With many new Illustrations. Two volumes.
Vol. I. Diseases of the Nerves and Spinal Cord. 3d Edition, Enlarged. Cloth, $4.00; Sheep, $5.00
Vol. II. Diseases of the Brain and Cranial Nerves; General and Functional Disease. 2d Edition. Cloth, $4.00; Sheep, $5.00
GOWERS. Syphilis and the Nervous System. $1.00

GOWERS. Clinical Lectures. A New Volume of Essays on the Diagnosis, Treatment, etc., of Diseases of the Nervous System. $2.00

GOWERS. Epilepsy and Other Chronic Convulsive Diseases. 2d Edition. *In Press.*

HORSLEY. The Brain and Spinal Cord. The Structure and Functions of. Numerous Illustrations. $2.50

ORMEROD. Diseases of the Nervous System. 66 Wood Engravings. $1.00

OSLER. Chorea and Choreiform Affections. $2.00

PERSHING. Diagnosis of Nervous and Mental Diseases. Illustrated. *In Press.*

PRESTON. Hysteria and Certain Allied Conditions. Their Nature and Treatment. Illustrated. $2.00

WOOD. Brain Work and Overwork. .40

NURSING (see also Massage).

Special Catalogue of Books for Nurses sent free upon application.

CANFIELD. Hygiene of the Sick-Room. A Book for Nurses and Others. Being a Brief Consideration of Asepsis, Antisepsis, Disinfection, Bacteriology, Immunity, Heating and Ventilation, and Kindred Subjects for the Use of Nurses and Other Intelligent Women. $1.25

CUFF. Lectures to Nurses on Medicine. New Edition. $1.25

DOMVILLE. Manual for Nurses and Others Engaged in Attending the Sick. 8th Edition. With Recipes for Sick-room Cookery, etc. .75

FULLERTON. Obstetric Nursing. 41 Ills. 5th Ed. $1.00

FULLERTON. Surgical Nursing. 3d Ed. 69 Ills. $1.00

GROFF. Materia Medica for Nurses. With Questions for Self-Examination and a very complete Glossary. $1.25

" It will undoubtedly prove a valuable aid to the nurse in securing a knowledge of drugs and their uses."—*The Medical Record, New York.*

HUMPHREY. A Manual for Nurses. Including General Anatomy and Physiology, Management of the Sick Room, etc. 17th Ed. Illustrated. $1.00

" In the fullest sense, Dr. Humphrey's book is a distinct advance on all previous manuals. It is, in point of fact, a concise treatise on medicine and surgery for the beginner, incorporating with the text the management of childbed and the hygiene of the sick-room. Its value is greatly enhanced by copious wood-cuts and diagrams of the bones and internal organs."—*British Medical Journal, London.*

STARR. The Hygiene of the Nursery. Including the General Regimen and Feeding of Infants and Children, and the Domestic Management of the Ordinary Emergencies of Early Life, Massage, etc. 6th Edition. 25 Illustrations. $1.00

TEMPERATURE AND CLINICAL CHARTS. See page 6.

VOSWINKEL. Surgical Nursing. Second Edition, Enlarged. 112 Illustrations. $1.00

OBSTETRICS.

CAZEAUX AND TARNIER. Midwifery. With Appendix by MUNDÉ. The Theory and Practice of Obstetrics, including the Diseases of Pregnancy and Parturition, Obstetrical Operations, etc. 8th Edition. Illustrated by Colored and other full-page Plates, and numerous Wood Engravings.   Cloth, $4.50 ; Full Leather, $5.50

EDGAR. Text-Book of Obstetrics. Illustrated.   *Preparing.*

FULLERTON. Obstetric Nursing. 5th Ed. Illustrated. $1.00

LANDIS. Compend of Obstetrics. 6th Edition, Revised by WM. H. WELLS, Assistant Demonstrator of Clinical Obstetrics, Jefferson Medical College. With 47 Illustrations, .80 ; Interleaved, $1.00

WINCKEL. Text-Book of Obstetrics, Including the Pathology and Therapeutics of the Puerperal State. Authorized Translation by J. CLIFTON EDGAR, M.D. Illus.   Cloth, $5.00

PATHOLOGY.

BARLOW. General Pathology. 795 pages. 8vo. $5.00

BLACK. Micro-Organisms. The Formation of Poisons.   .75

BLACKBURN. Autopsies. A Manual of Autopsies Designed for the Use of Hospitals for the Insane and other Public Institutions. Ten full-page Plates and other Illustrations. $1.25

CONN. Agricultural Bacteriology. Illustrated. *Nearly Ready.*

COPLIN. Manual of Pathology. Including Bacteriology, Technic of Post-Mortems, Methods of Pathologic Research, etc. 330 Illustrations, 7 Colored Plates. 3d Edition. $3.50

DA COSTA. Clinical Pathology of the Blood. Illus. *In Press.*

HEWLETT. Manual of Bacteriology. 75 Illustrations. $3.00

ROBERTS. Gynecological Pathology. Illus. *Nearly Ready.*

THAYER. Compend of General Pathology. Illustrated. *Nearly Ready.* .80 ; Interleaved, $1.00

THAYER. Compend of Special Pathology. Illustrated. *Nearly Ready.* .80 ; Interleaved, $1.00

VIRCHOW. Post-Mortem Examinations. 3d Edition.   .75

WHITACRE. Laboratory Text-Book of Pathology. With 121 Illustrations. $1.50

WILLIAMS. Bacteriology. A Manual for Students. 78 Illustrations. 2d Edition, Revised. *Just Ready.* $1.50

PHARMACY.

Special Catalogue of Books on Pharmacy sent free upon application.

COBLENTZ. Manual of Pharmacy. A Complete Text-Book by the Professor in the New York College of Pharmacy. 2d Edition, Revised and Enlarged. 437 Illus.   Cloth, $3.50 ; Sheep, $4.50

COBLENTZ. Volumetric Analysis. Illustrated.   *In Press.*

BEASLEY. Book of 3100 Prescriptions. Collected from the Practice of the Most Eminent Physicians and Surgeons—English, French, and American. A Compendious History of the Materia Medica, Lists of the Doses of all the Officinal and Established Preparations, an Index of Diseases and their Remedies. 7th Ed. $2.00

BEASLEY. Druggists' General Receipt Book. Comprising a Copious Veterinary Formulary, Recipes in Patent and Proprietary Medicines, Druggists' Nostrums, etc.; Perfumery and Cosmetics, Beverages, Dietetic Articles and Condiments, Trade Chemicals, Scientific Processes, and many Useful Tables. 10th Ed. $2.00

BEASLEY. Pharmaceutical Formulary. A Synopsis of the British, French, German, and United States Pharmacopœias. Comprising Standard and Approved Formulæ for the Preparations and Compounds Employed in Medicine. 12th Edition. $2.00

PROCTOR. Practical Pharmacy. Lectures on Practical Pharmacy. 3d Edition, with Illustrations and Elaborate Tables of Chemical Solubilities, etc. $3.00

ROBINSON. Latin Grammar of Pharmacy and Medicine. 3d Edition. With elaborate Vocabularies. $1.75

SAYRE. Organic Materia Medica and Pharmacognosy. An Introduction to the Study of the Vegetable Kingdom and the Vegetable and Animal Drugs. Comprising the Botanical and Physical Characteristics, Source, Constituents, and Pharmacopeial Preparations, Insects Injurious to Drugs, and Parmacal Botany. With sections on Histology and Microtechnique, by W. C. STEVENS. 374 Illustrations. Second Edition. Cloth, $4.50

SCOVILLE. The Art of Compounding. Second Edition, Revised and Enlarged. Cloth, $2.50

STEWART. Compend of Pharmacy. Based upon " Remington's Text-Book of Pharmacy." 5th Edition, Revised in Accordance with the U. S. Pharmacopœia, 1890. Complete Tables of Metric and English Weights and Measures. .80; Interleaved, $1.00

UNITED STATES PHARMACOPŒIA. 7th Decennial Revision. Cloth, $2.50 (postpaid, $2.77); Sheep, $3.00 (postpaid, $3.27); Interleaved, $4.00 (postpaid. $4.50); Printed on one side of page only, unbound, $3.50 (postpaid, $3.90).
Select Tables from the U. S. P. Being Nine of the Most Important and Useful Tables, Printed on Separate Sheets. Carefully put up in patent envelope. .25

POTTER. Hand-Book of Materia Medica, Pharmacy, and Therapeutics. 600 Prescriptions. 8th Ed. Clo., $5.00; Sh., $6.00

PHYSICAL DIAGNOSIS.

BROWN. Medical Diagnosis. A Manual of Clinical Methods. 4th Edition. 112 Illustrations. Cloth, $2.25

DA COSTA. Clinical Examination of the Blood. Illustrated. In Press.

FENWICK. Medical Diagnosis. 8th Edition. Rewritten and very much Enlarged. 135 Illustrations. Cloth, $2.50

MEMMINGER. Diagnosis by the Urine. 2d Ed. 24 Illus. $1.00

STEEL. Physical Signs of Pulmonary Disease. $1.25

TYSON. Hand-Book of Physical Diagnosis. For Students and Physicians. By the Professor of Clinical Medicine in the University of Pennsylvania. Illus. 3d Ed., Improved and Enlarged. With Colored and other Illustrations. $1.50

PHYSIOLOGY.

BIRCH. Practical Physiology. An Elementary Class Book. 62 Illustrations. $1.75

BRUBAKER. Compend of Physiology. 10th Edition, Revised and Enlarged. Illustrated. .80; Interleaved, $1.00

2

KIRKES. Physiology. (16th Authorized Edition. Dark-Red Cloth.) A Hand-Book of Physiology. 16th Edition, Revised, Rearranged, and Enlarged. By Prof. W. D. Halliburton, of Kings College, London. 671 Illustrations, some of which are printed in colors. Cloth, $3.00; Leather, $3.75

LANDOIS. A Text-Book of Human Physiology, Including Histology and Microscopical Anatomy, with Special Reference to the Requirements of Practical Medicine. 5th American, translated from the 9th German Edition, with Additions by Wm. Stirling, m.d.,d.sc. 845 Illus., many of which are printed in colors. *In Press.*

STARLING. Elements of Human Physiology. 100 Ills. $1.00

STIRLING. Outlines of Practical Physiology. Including Chemical and Experimental Physiology, with Special Reference to Practical Medicine. 3d Edition. 289 Illustrations. $2.00

TYSON. Cell Doctrine. Its History and Present State. $1.50

PRACTICE.

BEALE. On Slight Ailments; their Nature and Treatment. 2d Edition, Enlarged and Illustrated. $1.25

FAGGE. Practice of Medicine. 4th Edition, by P. H. Pye-Smith, m.d. 2 Volumes. *In Press.*

FOWLER. Dictionary of Practical Medicine. By various writers. An Encyclopædia of Medicine. Clo., $3.00; Half Mor. $4.00

GOULD AND PYLE. Cyclopedia of Practical Medicine and Surgery. A Concise Reference Handbook, Alphabetically Arranged, with particular Reference to Diagnosis and Treatment. Edited by Drs. Gould and Pyle, Assisted by 72 Special Contributors. Illustrated, one volume. Large Square Octavo, Uniform with "Gould's Illustrated Dictionary."
Sheep or Half Morocco, $10.00; with Thumb Index, $11.00
Half Russia, Thumb Index, $12 00

☞ *Complete descriptive circular free upon application.*

HUGHES. Compend of the Practice of Medicine. 6th Edition, Revised and Enlarged.
Part I. Continued, Eruptive, and Periodical Fevers, Diseases of the Stomach, Intestines, Peritoneum, Biliary Passages, Liver, Kidneys, etc., and General Diseases, etc.
Part II. Diseases of the Respiratory System, Circulatory System, and Nervous System; Diseases of the Blood, etc.
Price of each part, .80; Interleaved, $1.00
Physician's Edition. In one volume, including the above two parts, a Section on Skin Diseases, and an Index. 6th Revised Edition. 625 pp.   Full Morocco, Gilt Edge, $2.25

TAYLOR. Practice of Medicine. 5th Edition.   Cloth, $4.00

TYSON. The Practice of Medicine. By James Tyson, m.d., Professor of Medicine in the University of Pennsylvania. A Complete Systematic Text-book with Special Reference to Diagnosis and Treatment. 2d Edition, Enlarged and Revised. Colored Plates and 125 other Illustrations. 1222 Pages. Cloth, $5.50; Leather, $6.50

PRESCRIPTION BOOKS.

BEASLEY. Book of 3100 Prescriptions. Collected from the Practice of the Most Eminent Physicians and Surgeons—English, French, and American. A Compendious History of the Materia, Medica, Lists of the Doses of all Officinal and Established Preparations, and an Index of Diseases and their Remedies. 7th Ed. $2.00

BEASLEY. Druggists' General Receipt Book. Comprising a Copious Veterinary Formulary, Recipes in Patent and Proprietary Medicines, Druggists' Nostrums, etc. ; Perfumery and Cosmetics, Beverages, Dietetic Articles and Condiments, Trade Chemicals, Scientific Processes, and an Appendix of Useful Tables. 10th Edition, Revised. $2.00

BEASLEY. Pocket Formulary. A Synopsis of the British, French, German, and United States Pharmacopœias and the chief unofficial Formularies. 12th Edition. $2.00

SKIN.

BULKLEY. The Skin in Health and Disease. Illustrated. .40

CROCKER. Diseases of the Skin. Their Description, Pathology, Diagnosis, and Treatment, with Special Reference to the Skin Eruptions of Children. 92 Illus. 3d Edition. *Preparing.*

SCHAMBERG. Diseases of the Skin. 2d Edition, Revised and Enlarged. 105 Illustrations. Being No. 16 ? Quiz-Compend ? Series.
Cloth, .80 ; Interleaved, $1.00

VAN HARLINGEN. On Skin Diseases. A Practical Manual of Diagnosis and Treatment, with special reference to Differential Diagnosis. 3d Edition, Revised and Enlarged. With Formulæ and 60 Illustrations, some of which are printed in colors. $2.75

SURGERY AND SURGICAL DISEASES (see also Urinary Organs).

BERRY. Diseases of the Thyroid Gland and Their Surgical Treatment. Illustrated. *Just Ready.* $4.00

BUTLIN. Operative Surgery of Malignant Disease. 2d Edition. Illustrated. Octavo. $4.50

CRIPPS. Ovariotomy and Abdominal Surgery. Illus. $8.00

DEAVER. Surgical Anatomy. A Treatise on Human Anatomy in its Application to Medicine and Surgery. With about 400 very Handsome full-page Illustrations Engraved from Original Drawings made by special Artists from Dissections prepared for the purpose. Three Volumes. Royal Square Octavo.
Cloth, $21.00 ; Half Morocco or Sheep, $24.00 ; Half Russia, $27.00
Complete descriptive circular and special terms upon application.

DEAVER. Appendicitis, Its Symptoms, Diagnosis, Pathology, Treatment, and Complications. Elaborately Illustrated with Colored Plates and other Illustrations. 2d Edition. $3.50

DULLES. What to Do First in Accidents and Poisoning. 5th Edition. New Illustrations. $1.00

FULLERTON. Surgical Nursing. 3d Edition. 69 Illus. $1.00

HAMILTON. Lectures on Tumors. 3d Edition. $1.25

HEATH. Minor Surgery and Bandaging. 11th Ed., Revised and Enlarged. 176 Illustrations, Formulæ, Diet List, etc. $1.25

HEATH. Injuries and Diseases of the Jaws. 4th Ed. $4.50

HORWITZ. Compend of Surgery and Bandaging, including Minor Surgery, Amputations, Fractures, Dislocations, Surgical Diseases, and the Latest Antiseptic Rules, etc., with Differential Diagnosis and Treatment. 5th Edition, very much Enlarged and Rearranged. 167 Illustrations, 98 Formulæ. Clo., .80 ; Interleaved, $1.00

JACOBSON. Operations of Surgery. Over 200 Illustrations.
Cloth, $3.00; Leather, $4.00

KEHR. Gall-Stone Disease. Translated by WILLIAM WOTKYNS
SEYMOUR, M.D. *Just Ready.* $2.50

LANE. Surgery of the Head and Neck. 110 Illus. $5.00

MACREADY. A Treatise on Ruptures. 24 Full-page Litho-
graphed Plates and Numerous Wood Engravings.   Cloth, $6.00

MAKINS. Surgical Experiences in South Africa. 1899-1900.
Illustrated. *Just Ready.* $4.00

MAYLARD. Surgery of the Alimentary Canal. 97 Illustrations.
2d Edition, Revised. $3.00

MOULLIN. Text-Book of Surgery. With Special Reference to
Treatment. 3d American Edition. Revised and edited by JOHN B.
HAMILTON, M.D., LL.D., Professor of the Principles of Surgery and
Clinical Surgery, Rush Medical College, Chicago. 623 Illustrations,
many of which are printed in colors. Cloth, $6.00; Leather, $7.00

SMITH. Abdominal Surgery. Being a Systematic Description of
all the Principal Operations. 224 Illus. 6th Ed. 2 Vols. Clo., $10.00

SWAIN. Surgical Emergencies. Fifth Edition.   Cloth, $1.75

VOSWINKEL. Surgical Nursing. Second Edition, Revised and
Enlarged. 111 Illustrations. $1.00

WALSHAM. Manual of Practical Surgery. 7th Ed., Revised and Enlarged. 483 Engravings. 950 pages. $3.50

THROAT AND NOSE (see also Ear).

COHEN. The Throat and Voice. Illustrated.   .40

HALL. Diseases of the Nose and Throat. 2d Edition, Enlarged.
Two Colored Plates and 80 Illustrations. *Just Ready.* $2.75

HOLLOPETER. Hay Fever. Its Successful Treatment. $1.00

KNIGHT. Diseases of the Throat. A Manual for Students.
Illustrated.   *Nearly Ready.*

LAKE. Laryngeal Phthisis, or Consumption of the Throat.
Colored Illustrations. *Just Ready.* $2.00

MACKENZIE. Pharmacopœia of the London Hospital for
Dis. of the Throat. 5th Ed., Revised by Dr. F. G. HARVEY. $1.00

McBRIDE. Diseases of the Throat, Nose, and Ear. With colored Illustrations from original drawings. 3d Edition. $7.00

POTTER. Speech and its Defects. Considered Physiologically,
Pathologically, and Remedially. $1.00

SHEILD. Nasal Obstructions. Illustrated. *Just Ready.* $1.50

URINE AND URINARY ORGANS.

ACTON. The Functions and Disorders of the Reproductive
Organs in Childhood, Youth, Adult Age, and Advanced Life,
Considered in their Physiological, Social, and Moral Relations.
8th Edition. $1.75

BEALE. One Hundred Urinary Deposits. On eight sheets, for the Hospital, Laboratory, or Surgery. Paper, $2.00

HOLLAND. The Urine, the Gastric Contents, the Common Poisons, and the Milk. Memoranda, Chemical and Microscopical, for Laboratory Use. Illustrated and Interleaved. 6th Ed. $1.00

KLEEN. Diabetes and Glycosuria. $2.50

MEMMINGER. Diagnosis by the Urine. 2d Ed. 24 Illus. $1.00

MORRIS. Renal Surgery, with Special Reference to Stone in the Kidney and Ureter and to the Surgical Treatment of Calculous Anuria. Illustrated. $2.00.

MOULLIN. Enlargement of the Prostate. Its Treatment and Radical Cure. 2d Edition. Illustrated. $1.75

MOULLIN. Inflammation of the Bladder and Urinary Fever. Octavo. $1.50

SCOTT. The Urine. Its Clinical and Microscopical Examination. 41 Lithographic Plates and other Illustrations. Quarto. Cloth, $5.00

TYSON. Guide to Examination of the Urine. For the Use of Physicians and Students. With Colored Plate and Numerous Illustrations engraved on wood. 9th Edition, Revised. $1.25

VAN NUYS. Chemical Analysis of Urine. 39 Illus. $1.00

VENEREAL DISEASES.

COOPER. Syphilis. 2d Edition, Enlarged and Illustrated with 20 full-page Plates. $5.00

GOWERS. Syphilis and the Nervous System. 1.00

STURGIS AND CABOT. Student's Manual of Venereal Diseases. 7th Revised and Enlarged Ed 12mo. *Just Ready.* $1.25

VETERINARY.

BALLOU. Veterinary Anatomy and Physiology. 29 Graphic Illustrations. .80; Interleaved, $1.00

TUSON. Veterinary Pharmacopœia. Including the Outlines of Materia Medica and Therapeutics. 5th Edition. $2.25

WOMEN, DISEASES OF.

BISHOP. Uterine Fibromyomata. Their Pathology, Diagnosis, and Treatment. Illustrated. *Just Ready.* Cloth, $3 50

BYFORD (H. T.). Manual of Gynecology. Second Edition, Revised and Enlarged by 100 pages. 341 Illustrations. $3.00

DÜHRSSEN. A Manual of Gynecological Practice. 105 Illustrations. $1.50

FULLERTON. Surgical Nursing. 3d Edition, Revised and Enlarged. 69 Illustrations. $1.00

LEWERS. Diseases of Women. 146 Illus. 5th Ed. $2.50

MONTGOMERY. Practical Gynecology. A Complete Systematic Text-Book. 527 Illustrations. Cloth, $5.00; Leather, $6.00

ROBERTS. Gynecological Pathology. Illustrated.
Nearly Ready.

WELLS. Compend of Gynecology. Illustrated. 2d Edition.
.80; Interleaved, $1.00

COMPENDS.

From The Southern Clinic.

"We know of no series of books issued by any house that so fully meets our approval as these ? Quiz-Compends ?. They are well arranged, full, and concise, and are really the best line of text-books that could be found for either student or practitioner."

BLAKISTON'S ? QUIZ-COMPENDS?

The Best Series of Manuals for the Use of Students.

Price of each, Cloth, .80.   Interleaved, for taking Notes, $1.00.

☞ These Compends are based on the most popular text-books and the lectures of prominent professors, and are kept constantly revised, so that they may thoroughly represent the present state of the subjects upon which they treat.

☞ The authors have had large experience as Quiz-Masters and attaches of colleges, and are well acquainted with the wants of students.

☞ They are arranged in the most approved form, thorough and concise, containing over 600 fine illustrations, inserted wherever they could be used to advantage.

☞ Can be used by students of *any* college.

☞ They contain information nowhere else collected in such a condensed, practical shape. **Illustrated Circular free.**

No. 1. POTTER. HUMAN ANATOMY. Sixth Revised and Enlarged Edition. Including Visceral Anatomy. Can be used with either Morris's or Gray's Anatomy. 117 Illustrations and 16 Lithographic Plates of Nerves and Arteries, with Explanatory Tables, etc. By SAMUEL O. L. POTTER, M.D., Professor of the Practice of Medicine, College of Physicians and Surgeons, San Francisco; Brigade Surgeon, U. S. Vol.

No. 2. HUGHES. PRACTICE OF MEDICINE. Part I. Sixth Edition, Enlarged and Improved. By DANIEL E. HUGHES, M.D., Physician-in-Chief, Philadelphia Hospital, late Demonstrator of Clinical Medicine, Jefferson Medical College, Phila.

No. 3. HUGHES. PRACTICE OF MEDICINE. Part II. Sixth Edition, Revised and Improved. Same author as No. 2.

No. 4. BRUBAKER. PHYSIOLOGY. Tenth Edition, with Illustrations and a table of Physiological Constants. Enlarged and Revised. By A. P. BRUBAKER, M.D., Professor of Physiology and General Pathology in the Pennsylvania College of Dental Surgery; Adjunct Professor of Physiology, Jefferson Medical College, Philadelphia, etc.

No. 5. LANDIS. OBSTETRICS. Sixth Edition. By HENRY G. LANDIS, M.D. Revised and Edited by WM. H. WELLS, M.D., Instructor of Obstetrics, Jefferson Medical College, Philadelphia. Enlarged. 47 Illustrations.

No. 6. POTTER. MATERIA MEDICA, THERAPEUTICS, AND PRESCRIPTION WRITING. Sixth Revised Edition (U. S. P. 1890). By SAMUEL O. L. POTTER, M.D., Professor of Practice, College of Physicians and Surgeons, San Francisco; Brigade Surgeon, U. S. Vol.

? QUIZ-COMPENDS ?—Continued.

No. 7. WELLS. GYNECOLOGY. Second Edition. By Wm. H. Wells, m.d., Instructor of Obstetrics, Jefferson College, Philadelphia. 140 Illustrations.

No. 8. GOULD AND PYLE. DISEASES OF THE EYE AND REFRACTION. Second Edition. Including Treatment and Surgery, and a Section on Local Therapeutics. By George M. Gould, m.d., and W. L. Pyle, m.d. With Formulæ, Glossary, Tables, and 109 Illustrations, several of which are Colored.

No. 9. HORWITZ. SURGERY, Minor Surgery, and Bandaging. Fifth Edition, Enlarged and Improved. By Orville Horwitz, b. s., m.d., Clinical Professor of Genito-Urinary Surgery and Venereal Diseases in Jefferson Medical College; Surgeon to Philadelphia Hospital, etc. With 98 Formulæ and 71 Illustrations.

No. 10. LEFFMANN. MEDICAL CHEMISTRY. Fourth Edition. Including Urinalysis, Animal Chemistry, Chemistry of Milk, Blood, Tissues, the Secretions, etc. By Henry Leffmann, m.d., Professor of Chemistry in the Woman's Medical College of Penna; Pathological Chemist, Jefferson Medical College Hospital.

No. 11. STEWART. PHARMACY. Fifth Edition. Based upon Prof. Remington's Text-Book of Pharmacy. By F. E. Stewart, m.d., ph.g., late Quiz-Master in Pharmacy and Chemistry, Philadelphia College of Pharmacy; Lecturer at Jefferson Medical College. Carefully revised in accordance with the new U. S. P.

No. 12. BALLOU. VETERINARY ANATOMY AND PHYSIOLOGY. Illustrated. By Wm. R. Ballou, m.d., Professor of Equine Anatomy at New York College of Veterinary Surgeons; Physician to Bellevue Dispensary, etc. 29 graphic Illustrations

No. 13. WARREN. DENTAL PATHOLOGY AND DENTAL MEDICINE. Third Edition, Illustrated. Containing a Section on Emergencies. By Geo. W. Warren, d.d.s., Chief of Clinical Staff, Pennsylvania College of Dental Surgery.

No. 14. HATFIELD. DISEASES OF CHILDREN. Second Edition. Colored Plate. By Marcus P. Hatfield, Professor of Diseases of Children, Chicago Medical College.

No. 15. THAYER. GENERAL PATHOLOGY. By A. E. Thayer, m.d., Cornell University Medical College. Illustrated.

No. 16. SCHAMBERG. DISEASES OF THE SKIN. Second Edition. By Jay F. Schamberg, m.d., Professor of Diseases of the Skin, Philadelphia Polyclinic. Second Edition, Revised and Enlarged. 105 handsome Illustrations.

No. 17. CUSHING. HISTOLOGY. By H. H. Cushing, m.d., Demonstrator of Histology, Jefferson Medical College, Philadelphia. Illustrated.

No. 18. THAYER. SPECIAL PATHOLOGY. Illustrated. By same Author as No. 15.

Price, each, Cloth, .80.   Interleaved, for taking Notes, $1.00.

Careful attention has been given to the construction of each sentence, and while the books will be found to contain an immense amount of knowledge in small space, they will likewise be found easy reading; there is no stilted repetition of words; the style is clear, lucid, and distinct. The arrangement of subjects is systematic and thorough; there is a reason for every word. They contain over 600 illustrations.

Morris'
Anatomy

Second Edition, Revised and Enlarged.

790 Illustrations, of which many are in Colors.

Royal Octavo. Cloth, $6.00 ; Sheep, $7.00 ;
Half Russia, $8.00.
Special Thumb Index in Each Copy.

From The Medical Record, New York.

" The reproach that the English language can boast of no treatise on anatomy deserving to be ranked with the masterly works of Henle, Luschka, Hyrtl, and others, is fast losing its force.   During the past few years several works of great merit have appeared, and among these Morris's " Anatomy " seems destined to take first place in disputing the palm in anatomical fields with the German classics.   The nomenclature, arrangement, and entire general character resemble strongly those of the above-mentioned handbooks, while in the beauty and profuseness of its illustrations it surpasses them. . . . The ever-growing popularity of the book with teachers and students is an index of its value, and it may safely be recommended to all interested."

From The Philadelphia Medical Journal.

" Of all the text-books of moderate size on human anatomy in the English language, Morris is undoubtedly the most up-to-date and accurate."